Neuropilin: From Nervous System to Vascular and Tumor Biology

ADVANCES IN EXPERIMENTAL MEDICINE AND BIOLOGY

Recent Volumes in this Series

Volume 506
LACRIMAL GLAND, TEAR FILM, AND DRY EYE SYNDROME 3: BASIC SCIENCE AND CLINICAL RELEVANCE
Edited by David A. Sullivan, Michael E. Stern, Kazuo Tsubota, Darlene A. Dartt, Rose M. Sullivan, and B. Britt Bromberg

Volume 507
EICOSANOIDS AND OTHER BIOACTIVE LIPIDS IN CANCER, INFLAMMATION, AND RADIATION INJURY 5
Edited by Kenneth V. Honn, Lawrence J. Marnett, Santosh Nigam, and Charles Serhan

Volume 508
SENSORIMOTOR CONTROL OF MOVEMENT AND POSTURE
Edited by Simon C. Gandevia, Uwe Proske and Douglas G. Stuart

Volume 509
IRON CHELATION THERAPY
Edited by Chiam Hershko

Volume 510
OXYGEN TRANSPORT TO TISSUE XXIII: OXYGEN MEASUREMENTS IN THE 21ST CENTURY: BASIC TECHNIQUES AND CLINICAL RELEVANCE
Edited by David F. Wilson, John Biaglow and Anna Pastuszko

Volume 511
PEDIATRIC GENDER ASSIGNMENT: A CRITICAL REAPPRAISAL
Edited by Stephen A. Zderic, Douglas A. Canning, Michael C. Carr and Howard McC. Snyder III

Volume 512
LYMPHOCYTE ACTIVATION AND IMMUNE REGULATION IX: LYMPHOCYTE TRAFFIC AND HOMEOSTASIS
Edited by Sudhir Gupta, Eugene Butcher and William Paul

Volume 513
MOLECULAR AND CELLULAR BIOLOGY OF NEUROPROTECTION IN THE CNS
Edited by Christian Alzheimer

Volume 514
PHOTORECEPTORS AND CALCIUM
Edited by Wolfgang Baehr and Krzysztof Palczewski

Volume 515
NEUROPILIN: FROM NERVOUS SYSTEM TO VASCULAR AND TUMOR BIOLOGY
Edited by Dominique Bagnard

Volume 516
TRIPLE REPEAT DISORDERS OF THE NERVOUS SYSTEM
Edited by Lubov T. Timchenko

Volume 517
DOPAMINERGIC NEURON TRANSPLANTATION IN THE WEAVER MOUSE MODEL OF PARKINSON'S DISEASE
Edited by Lazaros C. Triarhou

Volume 518
ADVANCES IN MALE MEDIATED DEVELOPMENTAL TOXICITY
Edited by Bernard Robaire and Barbara F. Hales

A Continuation Order Plan is available for this series. A continuation order will bring delivery of each new volume immediately upon publication. Volumes are billed only upon actual shipment. For further information please contact the publisher.

Neuropilin: From Nervous System to Vascular and Tumor Biology

Edited by
Dominique Bagnard, Ph.D.
Maître de Conférences
Université Louis Pasteur
67084 Strasbourg, France
email: bagnard@neurochem.u-strasbg.fr

Kluwer Academic / Plenum Publishers
New York, Boston, Dordrecht, London, Moscow
Landes Bioscience / Eurekah.com
Georgetown, Texas, U.S.A.

Library of Congress Cataloging-in-Publication Data

CIP applied for but not received at time of publication.

Neuropilin: From Nervous System to Vascular and Tumor Biology
Edited by Dominique Bagnard
ISBN 0-306-47416-6
AEMB volume number: 515

Kluwer Academic / Plenum Publishers
233 Spring Street, New York, NY 10013
http://www.wkap.nl

Landes Bioscience
810 S. Church Street, Georgetown, TX 78626
http://www.landesbioscience.com; http://www.eurekah.com
Landes tracking number: 1-58706-168-6

10 9 8 7 6 5 4 3 2 1

A C.I.P. record for this book is available from the Library of Congress.

Printed in the United States of America.

PREFACE

Cell adhesion is one of the most important properties controlling embryonic development. Extremely precise cell-cell contacts are established according to the nature of adhesion molecules that are expressed on the cell surface. The identification of several families of adhesion molecules, well conserved throughout evolution, has been the basis of a considerable amount of work over the past 20 years that contributed to establish functions of cell adhesion in almost all organs. Nowadays, cell adhesion molecules are not just considered as cellular glue but are thought to play critical roles in cell signaling. Their ability to influence cell proliferation, migration, or differentiation depends on both cell surface adhesion properties and activation of intracellular pathways. The next challenge will be to understand how these molecules interact with each other to ensure specific functions in the morphogenesis of very sophisticated systems. Indeed, by exploring the cellular and molecular mechanisms of nervous system development, the group of H. Fujisawa in Japan identified in 1987 an adhesion molecule, neuropilin, highly expressed in the neuropile of amphibian optic tectum. Ten years later, two groups discovered that neuropilin is a receptor for guidance signals of the semaphorin family. Axon guidance is a critical step during brain development and the mechanisms ensuring growth cone navigation are beginning to be well understood. The semaphorins are bifunctional signals defining permissive or inhibitory pathways sensed by the growth cone. Moreover, a semaphorin can be repellent or attractive depending on the axonal populations. The complexity of the signaling cascade triggered by the semaphorin is further illustrated by the capacity of Sema3A to be repulsive for the axon and attractive for the dendrites of cortical neurons. Hence, an appropriate response of the growth cone requires the recruitment of a receptor complex enabling the integration of this varying information. The analysis of the structure of neuropilin revealed a very short intracellular domain lacking transduction capacities. Because of these works, several groups started to analyze the possible interactions of neuropilin and described various binding partners allowing semaphorin transduction. The current view considers neuropilin as the heart of a receptor complex consisting of multiple transmembrane molecules including tyrosine kinase receptors or other adhesion molecules. In front of the growing implication of neuropilin during various physiologic and pathophysiologic processes, we decided to edit this comprehensive book designed to illustrate the diverse functions of this basic adhesion molecule. The first part of the volume contains four Chapters presenting the discovery of neuropilin and demonstrating its principal functions in the nervous, vascular and immune systems. In the second part, four Chapters describe the molecular structure of neuropilin and dissect the mechanisms ensuring receptor complex formation with various molecules such as the Plexins, the Vascular Endothelial Growth Factor Receptors or other adhesion molecules such as L1. The last two Chapters focus on the pathophysiologic implication of neuropilin especially for tumor progression and nervous system lesions. More than an extensive description of a single molecule, this book proposes a general model for the understanding of a multi-functional factor, a model that may apply for a variety of signals. This volume illustrates how mechanisms are conserved in the development of various biological systems, from the nervous system, vascular system and immune system, how a single molecule is able to control extremely precise cell behavior through specific interactions, and finally how dysfunction of a particular signaling pathway may relate to disease. Understanding the functions ensured by such specific molecular interactions will certainly have broad implications for fundamental issues and clinical applications.

I would like to express my thanks to the authors who contributed in the production of this book by providing excellent reviews enriched by multiple useful figures. I would also like to thank R. Landes Bioscience and Kluwer Academic/Plenum Publishers for publishing the book.

Dominique Bagnard

PARTICIPANTS

Dominique Bagnard, Ph.D.
Maître de Conférences
Université Louis Pasteur
67084 Strasbourg, France

Dr. Elisabeth Brambilla
Laboratoire de Pathologie Cellulaire,
INSERM EMI, CHRU Grenoble
38043 Grenoble Cedex 09
France

Dr. Valérie Castellani
Laboratoire de Neurogenése et
Morphogenése dans
le Développement et chez l'Adulte
UMR CNRS 6156
Université de la Méditerranée
IBDM
Parc Scientifique de Luminy
13288 Marseille cedex 9
France
e-mail: castellani@lgpd.univ-mrs.fr

Dr. Fred De Winter
Graduate School for Neurosciences
Amsterdam
Netherlands Institute for Brain
Research
Meibergdreef 33
1105 AZ Amsterdam
The Netherlands

Dr. Harry Drabkin
University of Colorado Health
Sciences Center
Division of Medical Oncology,
Box B171
4200 East Ninth Avenue
Denver, CO 80262
USA

Dr. Hajime Fujisawa
Group of Developmental
Neurobiology
Division of Biological Science
Nagoya University Graduate
School of Science
Chikusa-ku, Nagoya 464-8602
Japan
e-mail: fujisawa@bio.nagoya-u.ac.jp

Dr. Yoshio Goshima
Department of Pharmacology
Yokohama City University School of
Medicine
3-9 Fukuura, Kanazawa-ku
Yokohama, Kanagawa 236-0004
Japan
e-mail: goshima@med.yokohama-
cu.ac.jp

Dr. Yael Herzog
Department of Biology, Technion
Israel Institute of Technology
Haifa, 32000
Israel

Dr. Anthony. J. G. D. Holtmaat
Graduate School for Neurosciences Amsterdam
Netherlands Institute for Brain Research
Meibergdreef 33
1105 AZ Amsterdam
The Netherlands

Dr. Ofra Kessler
Department of Biology, Technion
Israel Institute of Technology
Haifa, 32000
Israel

Dr. Michael Klagsbrun
Departments of Surgical Research and Pathology
Children's Hospital and Harvard Medical School
300 Longwood Avenue
Boston, MA 02115
USA
e-mail: klagsbrun@a1.tch.harvard.edu

Dr. Valérie Lemarchandel
Département d'Hématologie (U567), Institut Cochin
CNRS UMR 8104, Maternité Port-Royal
123 Boulevard de Port-Royal
75014 Paris
France

Dr. Roni Mamluk
Department of Surgical Research
Children's Hospital and Harvard Medical School
300 Longwood Avenue
Boston, MA 02115
USA

Dr. Fumio Nakamura
Department of Pharmacology
Yokohama City University School of Medicine
3-9 Fukuura, Kanazawa-ku
Yokohama, Kanagawa 236-0004
Japan

Dr. Gera Neufeld
Department of Biology, Technion
Israel Institute of Technology
Haifa, 32000
Israel
e-mail: gera@tx.technion.ac.il

Dr. Andreas Püschel
Institut für Allgemeine Zoologie und Genetik
Westfälische Wilhelms-Universität, Schloßplatz 5
48149 Münster
Germany
e-mail: apuschel@uni-muenster.de

Pr Joëlle Roche
IBMIG, Université de Poitiers
40 avenue du Recteur Pineau
86022 Poitiers Cedex
France
e-mail: joelle.roche@univ-poitiers.fr

Dr. Paul-Henri Roméo
Département d'Hématologie (U567), Institut Cochin
CNRS UMR 8104, Maternité Port-Royal
123 Boulevard de Port-Royal
75014 Paris
France
romeo@cochin.inserm.fr

Dr. Seiji Takashima
Internal Medicine and Therapeutics
Osaka University Graduate School of Medicine
Suita Osaka 565-0871
Japan

Dr. Marc Tessier-Lavigne
Department of Anatomy
Univ California San Francisco, Room S 1334
513 Parnassus Ave
San Francisco, CA 94143-0452
USA
e-mail: MARCTL@itsa.ucsf.edu

Dr. Rafaele Tordjman
Département d'Hématologie (U567), Institut Cochin
CNRS UMR 8104, Maternité Port-Royal
123 Boulevard de Port-Royal
75014 Paris
France

Dr. Joost Verhaagen
Graduate School for Neurosciences Amsterdam
Netherlands Institute for Brain Research
Meibergdreef 33
1105 AZ Amsterdam
The Netherlands
e-mail: J.Verhaagen@nih.knaw.nl

CONTENTS

COLOR PLATES xvii

1. FROM THE DISCOVERY OF NEUROPILIN
TO THE DETERMINATION OF ITS ADHESION SITES 1

Hajime Fujisawa

Summary 1
Introduction 1
Identification of Monoclonal Antibodies
that Recognize *Xenopus* NRP and Plex 2
Molecular Cloning and Structure of NRP 4
Expression of NRP in the Nervous System 4
Cell Adhesion Properties of NRP1 7
Conclusion 8

2. NEUROPILINS AS SEMAPHORIN RECEPTORS:
IN VIVO FUNCTIONS IN NEURONAL CELL MIGRATION
AND AXON GUIDANCE 13

Anil Bagri and Marc Tessier-Lavigne

Summary 13
Introduction 13
Identification and Characterization of Neuropilins as Semaphorin Receptors 14
In vivo Functions of Neuropilins in Nervous System Wiring
During Development 20
Conclusion 29

3. THE ROLE OF NEUROPILIN IN VASCULAR
AND TUMOR BIOLOGY 33

Michael Klagsbrun, Seiji Takashima and Roni Mamluk

Summary 33
Introduction 34
Neuropilin Expression in Endothelial Cells 36
Regulation of Neuropilin Expression in Blood Vessels 37
Neuropilin and Angiogenesis 37

Tumor Cell Neuropilin 39
Vascular Injury 41
Perspectives and Future Directions 43

4. NEUROPILIN-1 IN THE IMMUNE SYSTEM 49

Paul-Henri Romeo, Valérie Lemarchandel and Rafaele Tordjman

Summary 49
Introduction 49
Neuropilin-1 is Expressed by Dendritic Cells and Resting T cells 50
T Cell-Dendritic Cell Interaction Induces Neuropilin-1 Polarization in T Cells 50
Neuropilin-1 Promotes Cell-Cell Interactions 51
Neuropilin-1 Mediates the Dendritic Cells-Induced Proliferation of Resting T Cells 52
Discussion 52

5. STRUCTURAL AND FUNCTIONAL RELATION OF NEUROPILINS 55

Fumio Nakamura and Yoshio Goshima

Summary 55
Introduction 55
Primary Structure and Genomic Structure of Neuropilin 56
Binding Properties of NRP Domains 60
Neuropilin-1 Interacting Protein Binds to the Carboxyl Terminus of NRP1 63
Additional Receptor Required for Signal Transduction 63
Concluding Remarks 66

6. THE FUNCTION OF NEUROPILIN/PLEXIN COMPLEXES 71

Andreas W. Püschel

Summary 71
Introduction 71
Neuropilins Form the Ligand-Binding Subunit of the Sema3A Receptors 72
Plexins Act as the Signal-Transducing Subunit of Semaphorin Receptors 72
Plexins are Essential Components of the Sema3A Receptor 73
The Role of GTPases for Signal Transduction by Plexins 75
Open Questions 78

7. THE INTERACTION OF NEUROPILIN-1 AND NEUROPILIN-2 WITH TYROSINE-KINASE RECEPTORS FOR VEGF 81

Gera Neufeld, Ofra Kessler and Yael Herzog

Summary 81
Introduction 82

The Mechanism by Which NRP1 Affects VEGF Induced Signaling by the VEGFR2 Receptor 84
The Interaction of Neuropilins with VEGFR1 86
Conclusions 88

8. THE FUNCTION OF NEUROPILIN / L1 COMPLEX 91

V. Castellani

Summary 91
Introduction 91
L1 and NRP1 Associate Through Their Extracellular Domains 92
L1/NRP1 Complex Formation Regulates Axonal Responsiveness to a Secreted Semaphorin 93
L1/NRP1 Complex Specifies Growth Cone Responses to Sema3A 95
Soluble L1 Modulates Axonal Responsiveness to Sema3A 96
Other Putative Functions Served by L1/NRP1 Complex Formation 97
Pivotal Molecules in Axon Guidance 100

9. NEUROPILIN AND ITS LIGANDS IN NORMAL LUNG AND CANCER 103

Joëlle Roche, Harry Drabkin and Elisabeth Brambilla

Summary 103
Introduction 103
Neuropilin and Semaphorin in Normal Mice Lung Development 105
Neuropilins and Its Ligands in Human Lung Tumor 106

10. NEUROPILIN AND CLASS 3 SEMAPHORINS IN NERVOUS SYSTEM REGENERATION 115

Fred De Winter, Anthony J.G.D. Holtmaat and Joost Verhaagen

Summary 115
Introduction 116
General Features of CNS Regeneration 116
Semaphorin and Neuropilin in the Intact and Injured Olfactory System 117
Neuropilin Ligands are Expressed by the Fibroblast Component of Neural CNS Scars 119
Neuropilins are Expressed at the CNS Lesion Site 121
Neuropilin/Semaphorin Regulation in Rat Models for Status Epilepticus 126
General Aspects of PNS Regeneration 126
Neuropilin/Semaphorin Regulation in the Injured PNS 127
Conclusions 129

INDEX 141

ABBREVIATIONS

AB: Angular bundle
CA: Cornu Ammonis
CNS: Central nervous system
CRMP-2: Collapsin responsive mediator protein 2
CSPG: Chondroitin sulfated proteoglycans
CST: Cortico spinal tract
DG: Dentate gyrus
Dox: Doxycycline
DRG: Dorsal root ganglia
EC: Endothelial cell
EphB3: Ephrin B3
epl: External plexiform layer
GAP43: Growth associated protein-43
GAPs: Growth associated proteins
gl: Glomeruli layer
GPI: Glycosyl-phosphatidylinositol
HR: Hilar Region
HSPG: Heparan sulfate proteoglycans
HUVEC: Human umbilical vein Endothelial Cells
IgCAM: Cell adhesion molecule of the Ig superfamily
ipl: Inner plexiform layer
LOT: Lateral olfactory tract
MAb: Monoclonal antibody
MAG: Myeline associated glycoprotein
ml: Mitral cell Layer
ML: Molecular Layer
NRP: Neuropilin
NRP1: Neuropilin-1
NRP2: Neuropilin-2
onl: Olfactory nerve layer
ORN: Olfactory receptor neuron
Plex: Plexin
Plex-A1: Plexin A1
PLGF: Placenta growth factor
PlGF-2: Heparin binding form of PlGF
PlGF-2: Placenta growth factor-2
PNS: Peripheral nervous system

p-Sp: Para-aortic splanchnopleural mesoderm
ROP: Retinopathy of prematurity
RST: Rubro spinal tract
RTK: Receptor tyrosine kinase
RT-PCR: Reverse transcriptase polymerase chain reaction
SE: Status epileticus
Sema: Semaphorin
SG: Sympathetic ganglion
sNRP1: Soluble NRP1
Tet: Tetracycline
TLE: Temporal lobe epilepsy
TNF-α: tumor necrosis factor-α
TSC: Terminal Schwann cell
VEGF: Vascular endothelial growth factor
$VEGF_{121}$: 121 amino-acids long form of VEGF
$VEGF_{145}$: 145 amino-acids long form of VEGF
$VEGF_{165}$: 165 amino-acids long form of VEGF
VEGFR: Vascular endothelial growth factor receptor
vSMC: Vascular smooth muscle cells

COLOR PLATES

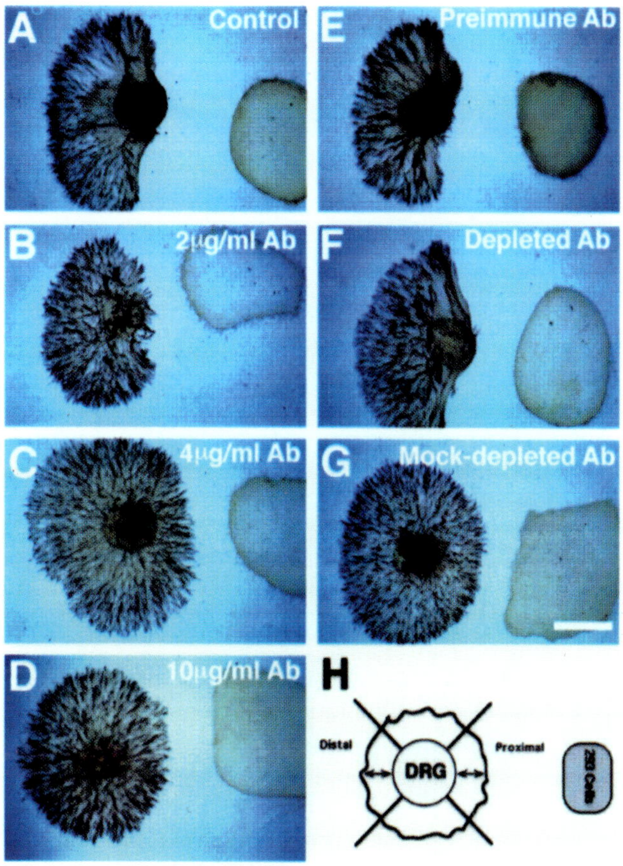

Figure C.1. Functional requirement of neuropilin-1 for Sema3 A-evoked repulsion of NGF-responsive DRG axons (reproduced from He and Tessier-Lavigne (1997)) as shown in Chapter 2, Figure 3.[15]

E14 rat DRG explants were cultured in collagen gels with 25 ng/ml NGF to elicit outgrowth of Sema III–responsive axons (Messersmith et al 1995). Explants were cocultured with aggregates of 293-EBNA cells secreting Sema3A–AP protein (right in each panel) in the presence of 0 μg/ml (A), 2μg/ml (B), 4 μg/ml (C), or 10 μg/ml (D) of anti-neuropilin IgG, 10 μg/ml of preimmune IgG (E), or 10 μg/ml of depleted (F) or mock-depleted (G) anti-neuropilin IgG for 40 hr. The explants were then fixed and visualized by wholemount immunostaining with the anti-neurofilament antibody NF-M. DRG neurites proximal to, but not distal to, the Sema3A–AP-secreting cells were repelled in the absence of anti-neuropolin antibody, an effect that was blocked in a dose-dependent fashion by addition of the antibody. (H) shows the procedure used to quantify the response. Scale bar: 350 μm.

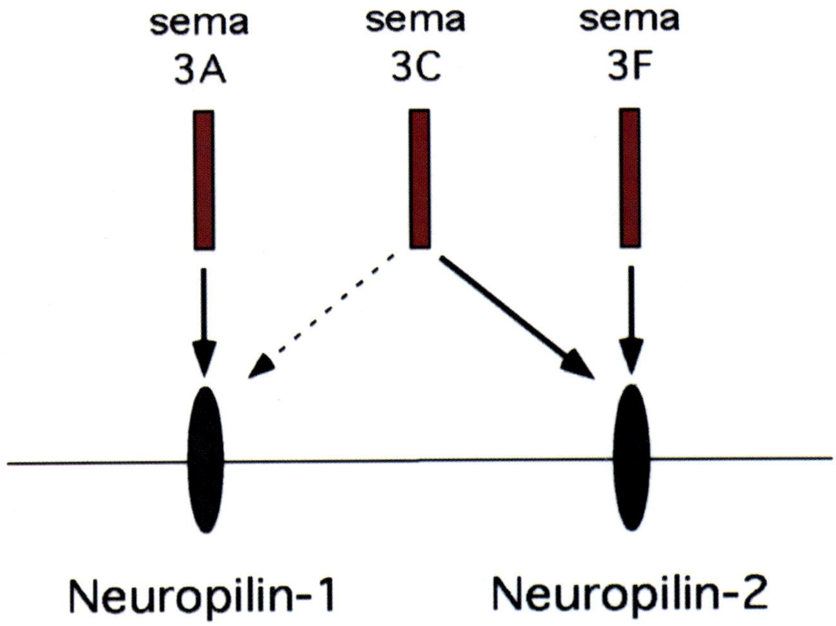

Figure C.2. Schematic representation of receptor specificity of different class 3 semaphorins as shown in Chapter 2, Figure 4.
Sema3A signaling is mediated via neuropilin-1, whereas sema3F signaling is mediated via neuropilin-2. Sema3C mediated signaling requires both neuropilins, but requires neuropilin-2 to a greater extent than neuropilin-1 (hence the dotted arrow).

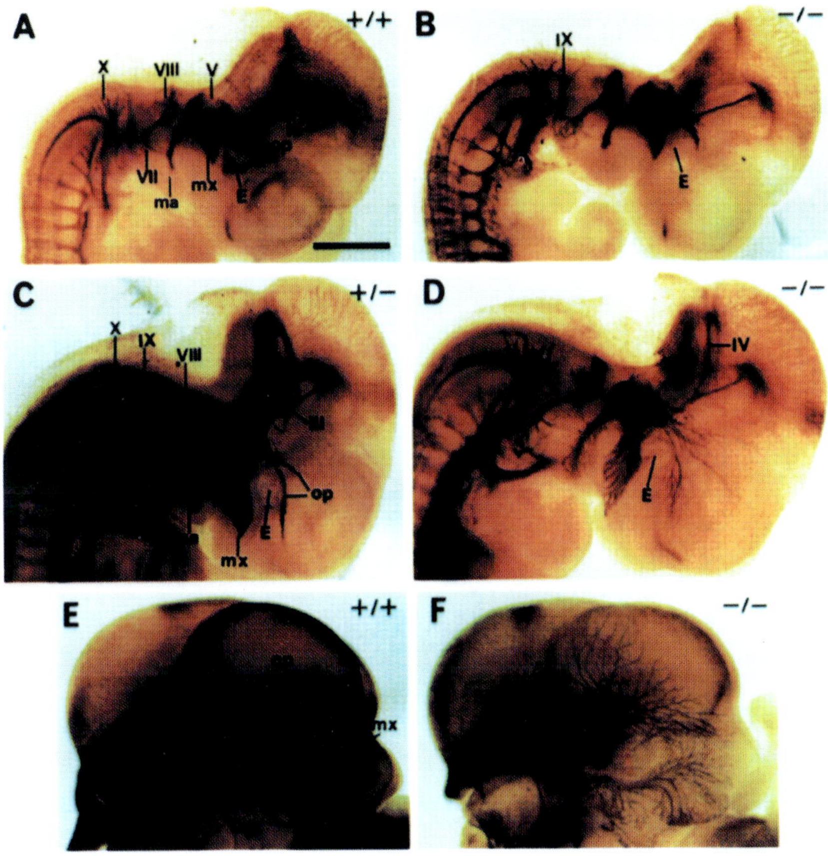

Figure C.3. Defects in projections of cranial nerves in neuropilin-1 mutant embryos (reproduced from Kitsukawa et al. 1997) as shown in Chapter 2, Figure5.[20]
Panels show whole-mount immunostaining with anti-neurofilament monoclonal antibody 2H3 of wildtype (+/+), heterozygous (+/–), and homozygous mutant (–/–) embryos at E9.5 (A and B), E10.5 (C and D), and E12.5 (E and F), to reveal defects in cranial nerves. III, oculomotor nerve; IV, trochlear nerve; V, trigeminal nerve; VII, facial nerve; VIII, vestibulocochlear nerve; IX, glossopharyngeal nerve; X, vagus nerve; op, ophthalmic nerve; mx, maxillary nerve; ma, mandibular nerve, E; eye. Scale bar, 1mm.

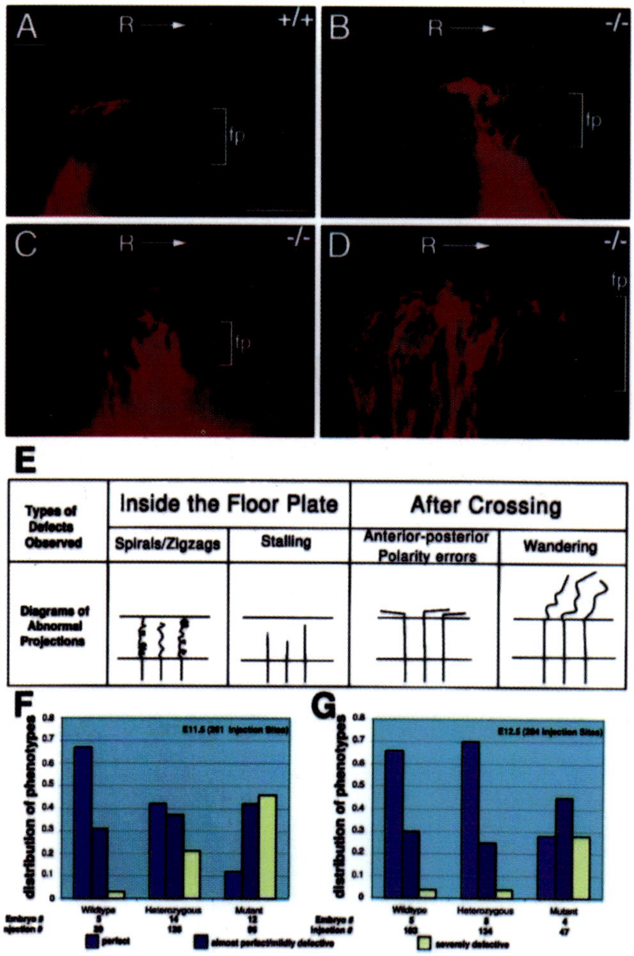

Figure C.4. Neuropilin-2 Is Required for Normal Midline Commissural Axon Pathfinding In Vivo in the Developing Spinal cord (reproduced from Zou et al, 2000) as shown in Chapter 2, Figure 7.[45]
(A–D) Visualization of commissural axon behavior at the floor plate (fp) in a wild-type E11.5 mouse embryo (A) and in three homozygous mutant neuropilin-2 E11.5 mouse embryos (B–D). Commissural axons are visualized following D:I injection in the dorsal spinal cord (off the bottom in each panel) in the "open book" configuration. Rostral (R) is to the right in each panel (indicated by arrow). In wild-type (A), commissural axons cross and turn rostrally in a very stereotyped fashion. A first example of pathfinding in a mutant embryo (B) shows randomization of the anterior-posterior projection patterns of commissural axons after exiting the floor plate, wavy axons and stalling growth cones inside the floor plate (note that the "waviness" starts approximately at the floor plate). A second example (C) shows commissural axons that are overshooting and wandering into the contralateral ventral spinal cord region after floor plate crossing. A third example (D) shows spiraling and wavy trajectories inside the floor plate (note again that the waviness is seen inside the floor plate, not before the floor plate). Scale bar: (A–C), 100 μm; (D), 66.7 μm, (E) Summary of commissural misrouting phenotypes in neuropilin-2 mutant mice. (F,G) Penetrance of defects in E11.5 and E12.5 embryos.

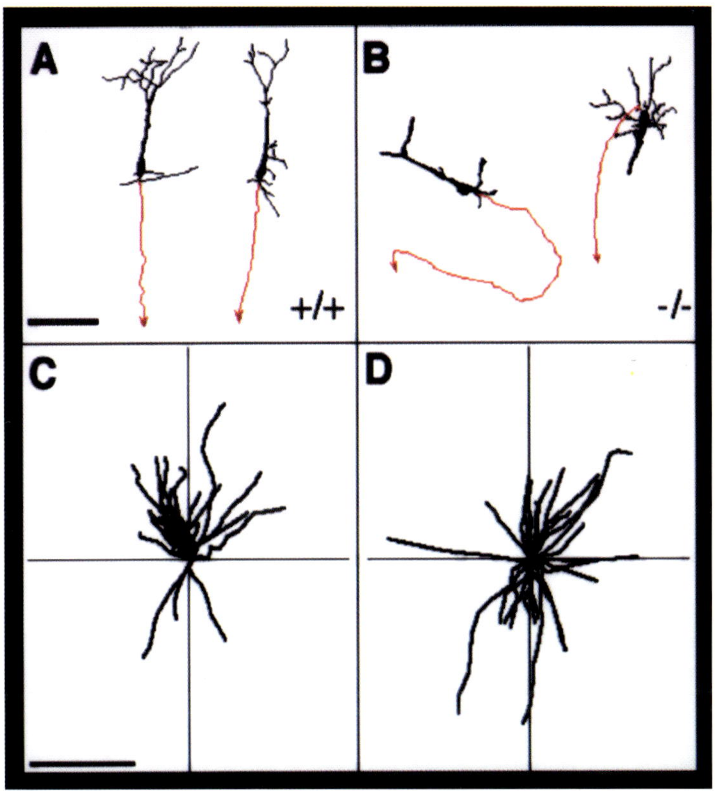

Figure C.5. Repulsion of axons and attraction of dendrites of cortical pyramidal neurons by Sema3A, activating neuropilin-1 (reproduced from Polleux et al, 1998; 2000) as shown in Chapter 2, Figure 8.[48,53]
Top panels: Evidence that endogenous Sema3A contributes to cortical axon guidance. Examples of the morphologies of cortical neurons in layers V-VI in wild-type (left) and Sema3A knock-out (right) mice labeled by white matter DiA injections. The pia is to the top; axons are indicated in red. Scale bar, 100 μm.
Bottom panels: Evidence that neuropilin-1 is required for correct apical dendrite orientation. Shown are apical dendrite orientation plots of E15 neurons from mice expressing GFP under the beta-actin promoter (and which express GFP) cultured on postnatal cortical slices in the absence (left) or presence (right) of a function-blocking antibody to neuropilin-1. The pia is to the top. Note that normally the dendrites are oriented towards the pia, but the anti-neuropilin-1 antibody interferes with this directionality. Scale bar, 200 μm.

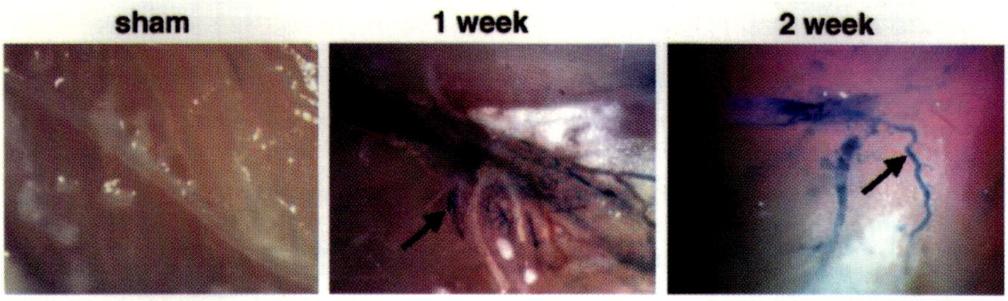

Figure C.6. NRP2-LacZ expression in an 8-10 week adult ischemic hindlimb as shown in Chapter 3, Figure 1.[91] Ischemia was induced by ligation of the femoral artery as previously described. Left: Two weeks after a sham operation without ligation of the femoral artery, there was very little if any NRP2 expression. Center: At one week after inducing ischemia, small sprouting capillaries surrounding the femoral artery expressed NRP2 at the site of the artery where vessel ligation occurred, (arrow). Right: At two weeks after inducing ischemia, NRP2 was expressed in mid-sized vessels growing from the cutting edge of the injured vessels and in small sprouting vessels (arrow).

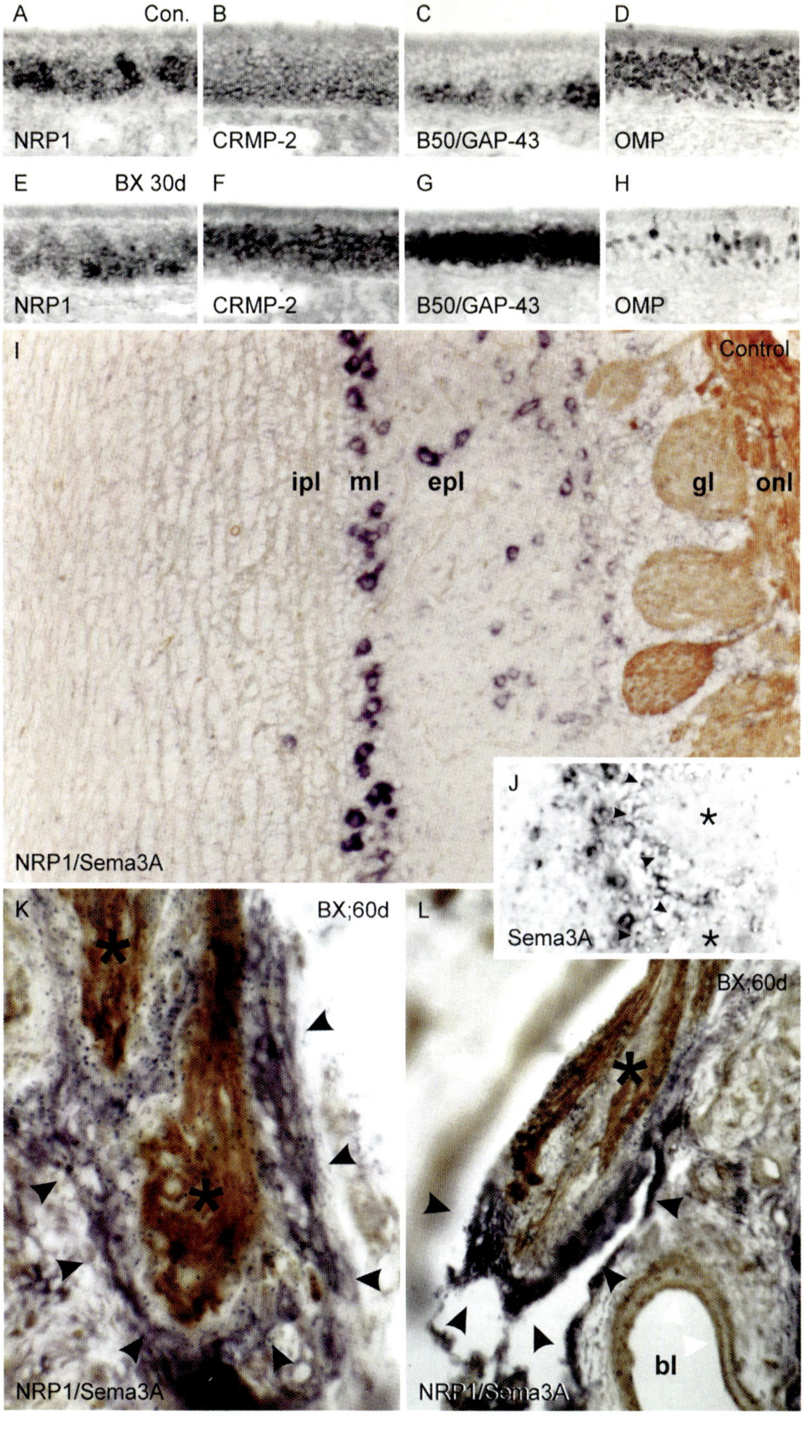
A
Con.
NRP1
B
CRMP-2
C
B50/GAP-43
D
OMP
E
BX 30d
NRP1
F
CRMP-2
G
B50/GAP-43
H
OMP
I
Control
ipl
ml
epl
gl
onl
NRP1/Sema3A
J
*
*
Sema3A
K
BX;60d
*
*
NRP1/Sema3A
L
BX;60d
*
bl
NRP1/Sema3A

Figure C.7. Neuropilin and semaphorin expression in the intact and injured olfactory system as shown in Chapter 10, Figure 1. (A-H) Horizontal sections through the olfactory epithelium of unlesioned adult rats (A-D) and bulbectomized rats (30 days post-lesion, E-F) were stained for NRP1 mRNA (A, E), CRMP-2 mRNA (B, F), B50/GAP-43 mRNA (C, G) and OMP protein (D, H). In control epithelium, NRP1 is expressed by small clusters of ORNs mainly located in the lower and middle parts of the epithelium (A), corresponding to immature B50/GAP43 positive neurons (C) and a subpopulation of mature OMP positive neurons respectively (D). CRMP-2 mRNA is highly expressed in the immature neurons located in the lower regions of the epithelium, but displays a declining gradient toward the apical surface (B). Both sustentacular cells and basal cells, located along the apical and the basal surface respectively, did not display any expression. Following bulbectomy the mature olfactory neurons degenerate and new neurons are formed from cells in the basal cell compartment of the olfactory epithelium. The number of immature B50/GAP-43 positive neurons increases dramatically (G), resulting in a enlarged overlap with the NRP1 (E) and CRMP-2 (F) expressing population. In contrast to the increase in immature neurons is the decrease in the number of OMP positive mature neurons (H), only a few neurons mature without a target. (I-L) Horizontal sections (rostral is to the top) of intact olfactory bulb (I) and 60 days following bulbectomy (K, L) were subjected to in situ hybridization for Sema3A (in purple) and immunocytochemistry for NRP1 (in brown). Mitral colis in the mitral cells layer (ml) and tufted cells located in the external plexiform layer (cpl) express high levels of Sema3A mRNA (I). Periglomerular cells, situated on the interface between the external plexiform layer and the glomeruli, express variable levels of Sema3A mRNA (black arrowheads in J). In the internal plexiform layer (ipl) no Sema3A expression was detected. Occasionally small cells in the olfactory nerve layer (onl) showed weak Sema3A expression. NRP1 positive ORN axons grow through the olfactory nerve layer (onl) and terminate in the glomeruli layer (gl) (I). Sixty days following removal of the olfactory bulb, clusters of Sema3A expressing cells (black arrow heads in K and L) encapsulate bundles of NRP1 positive nerve fibers (asterisks) entering the bulbar cavity. A NRP1 positive blood vessel (white arrowheads) in the bulbar cavity is indicated with Bl (L).

FROM THE DISCOVERY OF NEUROPILIN TO THE DETERMINATION OF ITS ADHESION SITES

Hajime Fujisawa

SUMMARY

Neuropilin (NRP) and plexin (Plex) that are now known to be semaphorin receptors were initially identified as antigens for monoclonal antibodies (MAbs) that bound to particular neuropiles and plexiform layers of the *Xenopus* tadpole optic tectum, several years before the discovery of semaphorin. The extracellular segment of the NRP protein is a mosaic of 3 functionally different protein motifs that are thought to be involved in molecular and/or cellular interactions, suggesting that NRP serves in a various cell-cell interaction by binding a variety of molecules. The first identified function of NRP was the cell adhesion activity; Cell reaggregation study using NRP-expressing cell lines revealed that NRP can mediate cell adhesion via heterophilic molecular interaction. Later, NRP was shown to bind semaphorins and vascular endothelial growth factor (VEGF). It was also shown that NRP makes receptor complexes with Plex to propagate semaphorin signals.

INTRODUCTION

Identification of molecules that guide axons with a high degree of precision is one of major subjects in developmental neurobiology. Over the past decade, a variety of axon guidance molecules with attractive or repulsive natures and their neuronal receptors have been identified.

Group of Developmental Neurobiology, Division of Biological Science, Nagoya University Graduate School of Science, Chikusa-ku, Nagoya 464-8602, Japan.

Semaphorins appear to function as repellents or attractants for neurons and regulate axonal growth. Since the discovery of first semaphorin, Sema3A (previously, collapsin-1), in 1993,[1] more than 20 semaphorins of secreted and transmembrane forms have been identified in various animal species. On the other hand, in 1997, a neuronal membrane protein referred to as neuropilin (NRP) was shown to bind Sema3A and propagate Sema3A-induced chemorepulsive signals into neurons.[2,3] Furthermore, in 1998, another neuronal membrane protein referred to as plexin (Plex) was shown to bind other semaphorins.[4,5] Nowadays, 2 NRPs and 10 Plexs have been identified and assumed to serve as receptors for semaphorins.

NRP and Plex, however, were discovered in 1987,[6] 6 years before the identification of Sema3A. Moreover, before 1997 when NRP was shown to be a semaphorin receptor, cell adhesion property was the only known function for NRP. In this Chapter, I will overview how NRP and Plex were discovered. In addition, I will describe cell adhesion activity of NRP and discuss its potential roles in neuronal development.

IDENTIFICATION OF MONOCLONAL ANTIBODIES THAT RECOGNIZE *XENOPUS* NRP AND PLEX

NPR and Plex were identified in the screening of molecules that would be involved in the establishment of the retinotectal projection in *Xenopus* tadpoles. The retinotectal projection system in lower vertebrates has been a good experimental model to elucidate mechanisms allowing specific neuronal connections. Developing and regenerating axons from different parts of the retina recognize discrete regions within the optic tectum to give raise to a fairly organized retinotopic neuronal connection. The chemoaffinity hypothesis, proposed by Sperry in 1963,[7] attributing neuronal recognition to specific cell surface labels is a prevailing idea. However, in the early 1980th, molecular mechanisms underlying specific neuronal recognition had remained obscure.

To isolate cell surface labels that play roles in specific neuronal connection between the retina and the optic tectum, we adopted hibridoma techniques. We immunized mice with dissociated *Xenopus* tadpole tectal cells, fused splenocytes with myeloma cells, and produced a panel of monoclonal antibodies (MAbs).[6] We performed immunostaining of tadpole optic tecta with supernatants of hibridoma cultures, and selected antibodies that bound to neuropiles or plexiform layers of the optic tectum and would recognize cell surface molecules. Among culture supernatants from more than 3,000 wells (through 10 fusions) we identified a monoclonal antibody (MAb) named as A5 (MAb-A5). The name of the antibody, A5, was derived from the well number of 96 well culture plate from which the hibridoma clone was isolated. The amphibian optic tectum has a laminar structure, defining layers 1 to 9. MAb-A5 preferentially bound to the most superficial neuropile (the 8th and 9th layers) that is the termination site of retinal axons (the optic nerve) (Fig. 1A). The binding of MAb-A5 was diminished by treatment of sections of living optic tectum with trypsin, suggesting that the antigen recognizes cell surface proteins. MAb-A5

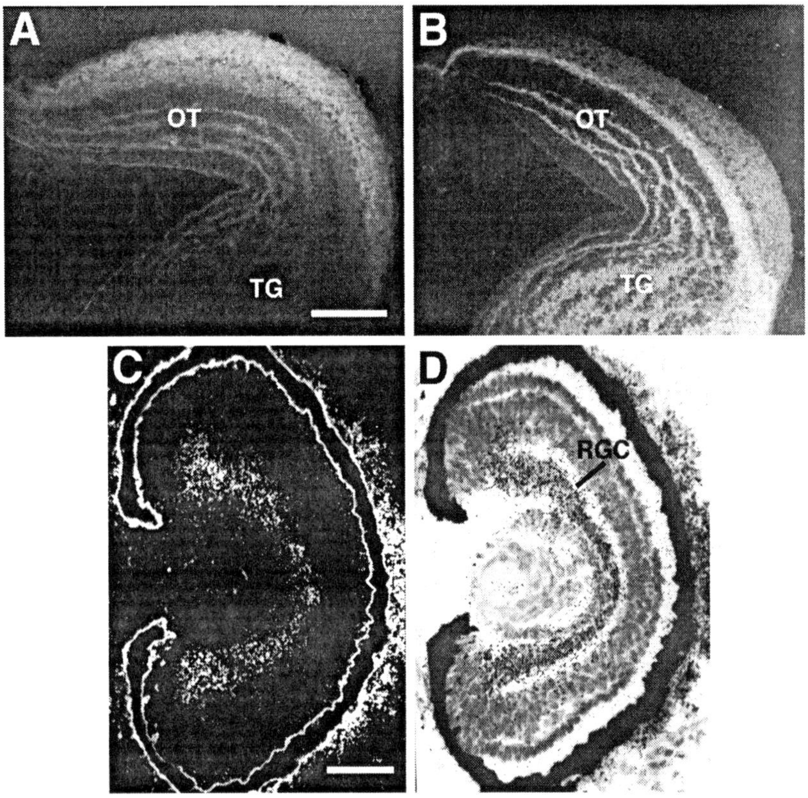

Figure 1. Binding of MAb-A5 and MAb-B2 to the optic tectum and expression of the antigen for MAb-A5 (*Xenopus* NRP1) in the neural retina
A, B: Immunofluoresence of MAb-A5 (A) and MAb-B2 (B) on sections of the optic tectum (OT) of *Xenopus* tadpoles at stage 53. The binding of MAb-A5 is restricted to the superficial neuropile, while MAb-B2 to the deeper plexiform layers of the optic tectum and the tegmentum (TG). C, D: Expression of NRP1 transcripts in the neural retina of *Xenopus* embryos at stage 41 detected by *in situ* hybridization; dark-field (C) and bright field (D) images of the same section. NRP1 is restrictively expressed in retinal ganglion cells (RGC). Scale bar (in A), 200 μm for A, B; (in C) 50 μm for C, D.

immuno-adsorbed a single polypeptide with an apparent molecular mass of 140 kDa. Later, in situ hybridization analysis and immunohistochemistry showed that the antigen for MAb-A5 is expressed in retinal ganglion cells that give raise to retinal axons (Fig. 1C,D, also see reference 8), as well as tectal neurons.

Interestingly, in the same fusion, we isolated another MAb named as B2 (MAb-B2).[6] In contrast to MAb-A5, MAb-B2 bound to plexiform layers in the deeper part of the optic tectum (Fig. 1B). The overall binding patterns for MAb-A5 and MAb-B2 in the optic tectum was apparently complementary. The antigen recognized by MAb-B2 was a peptide with a molecular mass of 200-220 kDa.

Based on the preferential binding of MAb-A5 to the neuropile, the antigen for MAb-A5 was named as neuropilin (NRP). On the other hand, the antigen for MAb-B2 was named as plexin (Plex),[9] a molecule expressed in the plexiform layers of the optic tectum and the neural retina[10].

MOLECULAR CLONING AND STRUCTURE OF NRP

Both MAb-A5 and MAb-B2 were not adequate for the screening of expression library. Therefore, we affinity purified the antigens for MAb-A5 and MAb-B2 by immuno-adsorption from more than 50,000 *Xenopus* tadpole brains, immunized rats with the antigens, and obtained A5-specific and B2-specific antisera. By using the antisera, we screened λgt11 expression library prepared from tadpole brain mRNAs. The cDNA cloning revealed that both the antigens for MAb-A5 (NRP) and MAb-B2 (Plex) were type 1 membrane glycoproteins.[9,11] Nowadays, NRP homologues have been isolated in various vertebrate species, including chicken,[12] mouse,[13] rat and human,[2,3] but not in invertebrates. As another NRP-related molecule has been identified,[3,14] the original NRP is referred to as neuropilin-1 (NRP1), and the new one as neuropilin-2 (NRP2). The primary structure of NRP1 is highly conserved among vertebrate species. For example, overall amino acid identity is 74.4% between the *Xenopus* and chick NRP1, and 72.6% between *Xenopus* and mouse NRP1. On the other hand, several Plex-related molecules have been identified in both invertebrates and vertebrates,[4,5,15-18] and are grouped into 4 subfamilies, PlexA, -B, -C and -D subfamilies. The original Plex found in *Xenopus* belongs to the PlexA subfamily (*Xenopus* PlexA1).[17]

As depicted in Figure 2, the extracellular part of NRP1 and NRP2 is composed of 3 unique domains referred to as a1/a2, b1/b2, and c, which are shared by a wide variety of molecules.[11-13] The a1/a2 domains have striking similarities to a motif found in the complement components C1r and C1s, bone morphogenetic protein-1 (BMP-1) and the *Drosophila* dorsal-ventral patterning protein Tolloid. The a1/a2-like motifs in these molecules have been assumed to be involved in molecular interaction. A motif similar to the b1/b2 domains of the NRP protein has been found in the coagulation factors V and VIII, and the extracellular part of a receptor tyrosine kinase discoidin domain receptor (DDR) all of which are expected to play roles in interaction with cell surfaces. The central portion of the c domain coincides with a module designated as the MAM domain which is contained in such functionally diverse proteins as the receptor protein tyrosine phosphatase and the metalloendopeptidases meprins, proteins that have been suggested to display adhesive functions.

EXPRESSION OF NRP IN THE NERVOUS SYSTEM

Immunohistochemical and in situ hybridization analyses performed on various vertebrate species have clarified the general features of NRP-expression.[6,8,11-14,19-23]

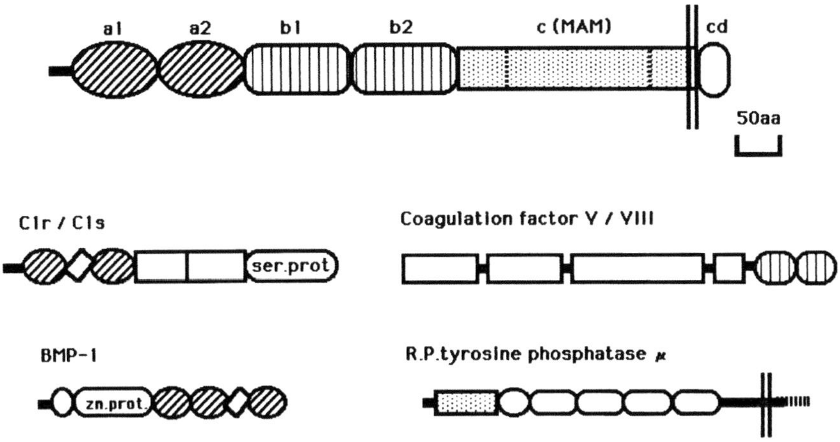

Figure 2. Primary structures of NRP and related molecules
Cd: cytoplasmic domain; ser.prot. serine protease domain; zn.prot. zinc protease domain.

First, the expression of NRP is limited to particular classes of neurons. Most peripheral sensory and autonomic ganglia, motor neuron pools in the spinal cord and the motor nuclei in the medulla, neurons in the hippocampal formation, cortical neurons, retinal ganglion cells, olfactory receptors and their central targets are the major sites for the NRP1-expression. Interestingly, NRP1 is expressed in retinal ganglion cells of *Xenopus* embryos and tadpoles[8,11] and mouse embryos[13] but not chick embryos.[12] The lack of NRP1-expression in chick retinal ganglion cells provided a base for the ectopic expression of NRP1 using a viral promoter in these cells to test functions of NRP1.[24] The expression patterns of NRP2 in the nervous systems are partially overlapped but mostly complementary to that of NRP1.[14] For example, in the mouse olfactory system, NRP1 is mainly expressed in the principal olfactory pathway while NRP2 is found in the accessory olfactory pathway.

Second, the expression of NRP in nervous systems is developmentally regulated. Both NRP1 and NRP2 are strongly expressed in developing but not adult nervous tissue, except the olfactory epithelium and the hippocampus where replacement of neurons occurs even in the adults. In both the peripheral and central nervous systems, NRP1 begins to appear in newly differentiated neurons, persists throughout the period in which axonal growth is active, and then diminishes after the frameworks of neuronal circuits have been accomplished. A good example for the axonal growth-associated expression of NRP1 is the regeneration of the optic nerve in *Xenopus*. The expression of NRP1 in the optic nerve is strong in embryos, but almost null in tadpoles after stage 50. When the tadpole optic nerves are crushed and prompted to regenerate, the NRP1 proteins reappear in the regenerating optic nerve fibers, persist during the following few weeks, and decline once the retinotectal projection is re-established.[8]

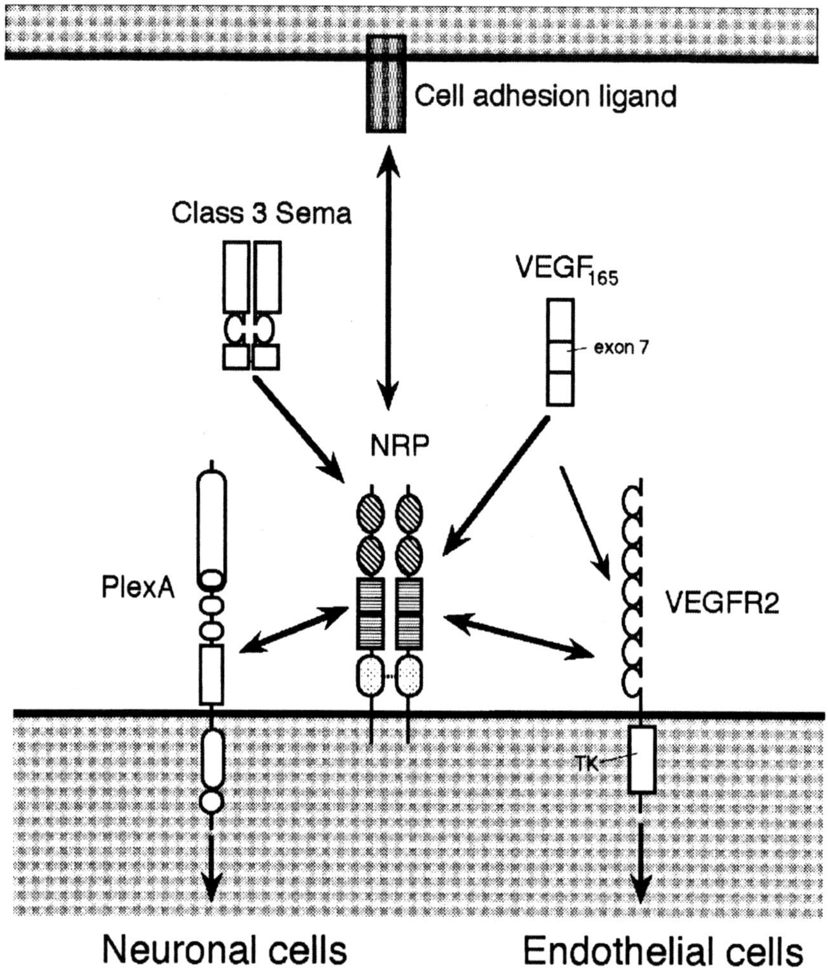

Figure 3. Multiple functions of NRP
In neuronal cells, NRP makes receptor complexes with members of the PlexA subfamily (PlexA) and propagates signals of secreted semaphorins of the class 3 (Class 3 Sema). In endothelial cells, NRP functions as coreceptor for a VEGF receptor, VEGFR2, and propagates signals of $VEGF_{165}$. NRP also interacts with unknown molecules (Cell adhesion ligand) of other cells to mediate cell adhesion. TK; tyrosine kinase domain.

The developmentally regulated expression of NRP in the nervous systems has suggested that the molecule plays some roles in neural development. Since the discovery of *Xenopus* NRP1, several approaches have been attempted to clarify functions of NRP and now shown that NRP can interact with secreted semaphorins of the class 3 to mediate semaphoring-induced chemorepulsive signals[2,3] (Fig. 3) and regulate axon guidance and nerve fiber patterning in developing mouse embryos.[25-28]

The differential expression of NRP1 and NRP2 provide anatomical bases for different sensitivity of these neurons to the class 3 semaphorins[14] and different neuronal phenotypes between the *NRP1*[25] and *NRP2*[27,28] mutant mice that had been produced by targeted disruption of the *NRP1* and *NRP2* genes. Though the functions of NRP1 in the *Xenopus* retinotectal projection system had been obscured for a long time, a recent study by Campbell et al[29] shows that NRP1-mediated Sema3A signals play roles in the guidance of embryonic retinal axons. To our surprise, it has been shown that NRPs interact with members of the PlexA subfamily to make receptor complexes for semaphorins of the class 3 (Fig. 3).[18,30,31] As 3 members of the PlexA subfamily are expressed in developing nervous systems in diverse patterns,[32] combination of NRPs and Plexs in given neurons may serve as semaphorin receptors and induce a diverse array of behaviors in axons to establish stereotyped patterns of neuron networks.

In addition to nervous systems, NRP is also expressed in endothelial cells[12,20] and function as a coreceptor for the vascular endothelial growth factor (VEGF) receptor, VEGFR2 (Flk-1/KDR), to mediate signals of $VEGF_{165}$ (an isoform of VEGF that contains a domain encoded by the exon 7 of the *VEGF* gene; see Fig. 3)[33] and regulate embryonic vessel formation.[20,34]

CELL ADHESION PROPERTIES OF NRP1

NRP serves as cell adhesion receptors, as well as receptors for semaphorins.

To examine cell adhesion activity of NRP1, we introduced chick or mouse NRP1 cDNAs into a mouse fibroblastic cell line (L cells), isolated cells that stably expressed NRP1, and then performed a cell aggregation assay.[12] The parental L cells had no aggregability by themselves without Ca^{2+} or Mg^{2+}. On the contrary, the NRP1-expressing L cells showed the ability to aggregate. When a mixture of the parental L cells and NRP1-expressing L cells was reaggregated, the parental L cells were incorporated into the aggregates (Fig. 4A-C), suggesting that NRP1 mediates cell adhesion by interactions with molecules expressed on cell surfaces of L cells. Pretreatment of L cells with trypsin abolished the incorporation of the cells into aggregates,[12] indicating that cell adhesion ligands for NRP1 are protease-sensitive molecules.

Structural and functional analysis on NRP1 has shown that members of the class 3 semaphorin can bind to the a1/a2 and b1/b2 domains of NRP1,[23,24] and $VEGF_{165}$ to the b1/b2 domains.[23] Moreover, NRP1 can physically interact with the members of PlexA subfamily.[18,30,31] Therefore, we determined cell adhesion sites of the NRP1 protein to examine whether cell adhesion properties of NRP1 is independent to these known NRP1 functions.[35] We produced cell lines expressing mutant NRP1s in which the extracellular domains were deleted in various combinations, and tested their cell adhesion activity. The cell aggregation analyses showed that the b1/b2 but not a1/a2 or c domains were essential to the cell adhesion activity of NRP1. As L cells bound to recombinant protein for the b1 and b2 domains, these 2 domains were expected to mediate cell adhesion independently. Then, we produced

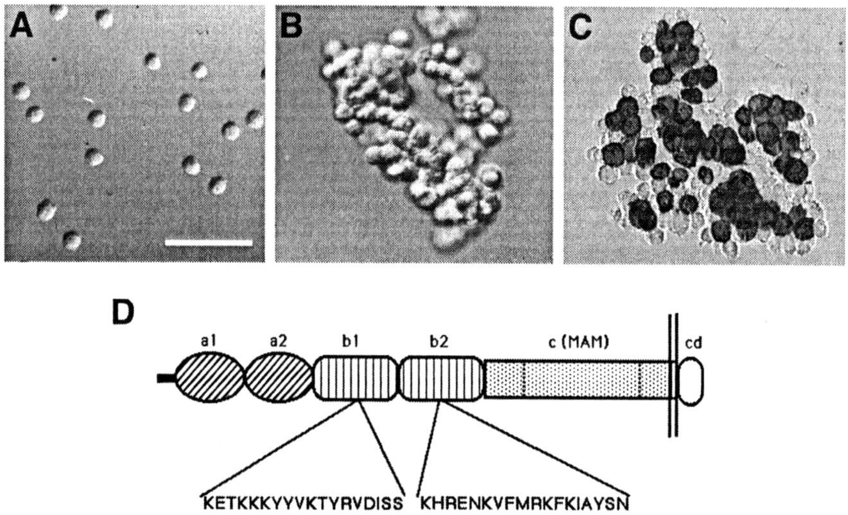

Figure 4. Cell adhesion properties of NRP1
A-C: Cell reaggregation assay on parental L cells (A), L cells expressing mouse NRP1 (mNRP1) (B), and a mixture of the parental L cells and mNRP1-expressing L cells (C); phase microscopy (A, B) and immunostaining with anti-mNRP1 antibody (C). D: Amino acid sequences of the cell adhesion sites within the b1 and b2 domains of the mNRP1 protein. Scale bar (in A), 100 μm for A-C.

a variety of recombinant proteins for the b1 and b2 domains and tested their cell adhesion activities. We determined the adhesion sites within an 18 amino acid stretch in the central part of these domains that are essential for the cell adhesion activity of NRP1 (Fig. 4D). Members of the class 3 semaphorin (Sema3A, Sema3B and Sema3C) or PlexA subfamily (PlexA1, -A2 and -A3) did not interact with recombinant proteins for the cell adhesion site of NRP1. In addition, $VEGF_{165}$ did not interfere the NRP1-mediated cell adhesion. These results indicate that the cell adhesion sites of NRP1 differ to the interaction sites for Sema3A, VEGF or Plex.

The cell adhesion sites within the b1 and b2 domains are conserved among all NRP1s from different vertebrate species, suggesting that cell adhesion activity is a universal function of NRP1. As the amino acid sequences of the cell adhesion sites of NRP1 do not closely resemble the corresponding regions of NRP2, it is open to question whether NRP2 can mediate cell adhesion as NRP1 does.

CONCLUSION

The cell transfection studies clearly demonstrate a cell adhesion activity of NRP. The question is how and which steps of neural development the cell adhesion activity of NRP1 regulates.

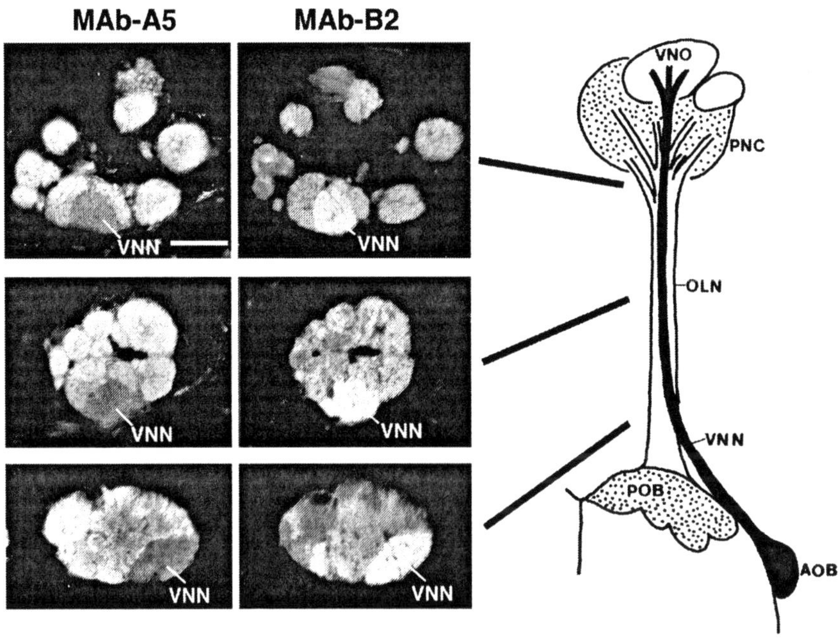

Figure 5. Pathway segregation of olfactory axons in *Xenopus* tadpoles
Adjacent sections of the olfactory nerve (OLN) and vomeronasal nerve (VNN) made at various levels from the nose to the olfactory bulb were immunostained with MAb-A5 and MAb-B2 that specifically recognize NRP1 and PlexA1, respectively (immunofluorescent staining). The vomeronasal nerve expresses PlexA1 but not NRP1. Note that MAb-A5-positive and MAb-B2-positive olfactory axons are almost evenly mixed at the proximal level of the olfactory nerve, but segregated at the distal end of the nerve. PNC: the principal nasal cavity; VNO: the vomeronasal organ; POB: the principal olfactory bulb; AOB: the accessory olfactory bulb. Scale bar, 100 μm.

Several lines of study carried out on the *Xenopus* and mouse nervous systems have suggested the involvement of NRP1 in nerve fiber fasciculation and aggregation of neural cells. In the *Xenopus*, the principal olfactory receptors are divided into at least 2 subclasses by virtue of the expression levels of NRP1 and PlexA1, the NRP-predominant receptors that express high levels of NRP1 and low levels of the PlexA1, and the Plex-predominant receptors that express high levels of PlexA1 and low levels of NRP1. These olfactory receptor subclasses are evenly distributed within the olfactory epithelium, and their axons (olfactory axons) are initially intermingled with each other. However, the NRP-predominant and the Plex-predominant olfactory axon subclasses become gradually segregated throughout their courses from the nose to the cerebrum, and eventually become completely separated and project to specified glomeruli in topographically related regions within the main olfactory bulb (Fig. 5; also see ref. 36). The sorting of olfactory axon subclasses within the olfactory nerve cannot simply be explained by chemorepulsive functions of semaphorins, rather might be explained by the cell adhesion activity of NRP1; NRP1

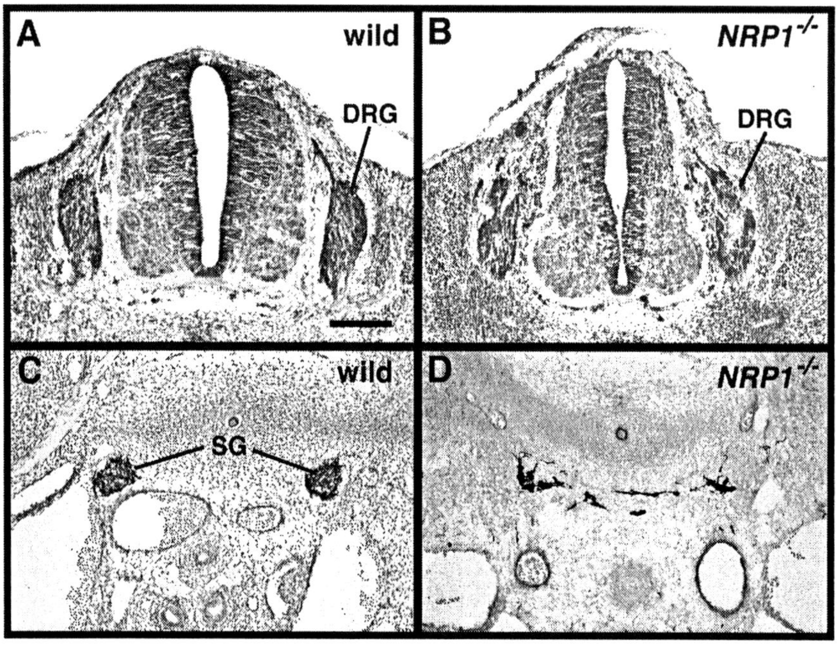

Figure 6. Morphology of peripheral ganglia in the *NRP1* mutant embryos
A, B: The dorsal root ganglia (DRG) of the wild-type and *NRP1* mutant (*NRP1*$^{-/-}$) mouse embryos at E12.5. Sections were stained with Hematoxylin-Eosin. C, D: The sympathetic ganglia (SG) of the wild-type and *NRP1* mutant (*NRP1*$^{-/-}$) mouse embryos at E12.5, immunostained with anti-TH antibody. Scale bar, (in A) 200 μm for A-D.

probably plays a role in axon-axon contact by interacting with adhesion ligands on axons.

On the other hand, it has been shown that, in the *NRP1* mutant embryos, cell packaging in the dorsal root ganglia (DRGs) were loose (Fig. 6A,B; also see ref. 25), and sympathetic ganglion (SG) neurons failed to be aggregated into ganglia but were displaced (Fig. 6C,D; also see ref. 37). As the regression in cell packaging in DRGs and SGs was also observed in the *Sema3A* mutant embryos,[37,38] Sema3A expressed in the tissues surrounding DRGs and SGs effects on neural cell aggregation. It is open to question how Sema3A promotes neuronal cell aggregability. One possibility is that Sema3A up-regulates NRP1 expression in these neurons to increases cell adhesiveness. More recently, NRP1 has been shown to form a complex with a neuronal cell adhesion molecule, L1.[39] Therefore, it is also likely that Sema3A modifies the interaction of NRP1 with L1 or other cell adhesion molecules and increases cell adhesiveness.

Much attention has been given on NRP functions as semaphorin receptor and VEGF receptor, but few on its function as cell adhesion receptor. The above evidences are still circumstantial to establish the functions of cell adhesion activity of

NRP in neural development, requiring further analyses, in particular, the identification of cell adhesion ligands for NRP1.

ACKNOWLEDGMENTS

This work was funded by grants from the CREST (Core Research for Evolutional Science and Technology) of Japan Science and Technology Corporation (JST) and the Japan Society for Promotion of Science.

REFERENCES

1. Luo Y, Raible D, Raper JA. Collapsin: A protein in brain that induces the collapse and paralysis of neuronal growth cones. Cell 1993; 75:217-227.
2. He Z, Tessier-Lavigne M. Neuropilin is a receptor for the axonal chemorepellent semaphorin III. Cell 1997; 90:739-751.
3. Kolodkin AL, Levengood DV, Rowe EG et al. Neuropilin is a semaphorin III receptor. Cell 1997; 90:753-762.
4. Comeau MR, Johnson R, DuBose RF et al. A poxvirus-encoded semaphorin induces cytokine production from monocytes and binds to a novel cellular semaphorin receptor, VESPR. Immunity 1998; 8:473-482.
5. Winberg ML, Noordermeer JN, Tamagnone L et al. Plexin A is a neuronal semaphorin receptor that controls axon guidance. Cell 1998; 95:903-916.
6. Takagi S, Tsuji T, Amagai T et al. Specific cell surface labels in the visual centers of *Xenopus laevis* tadpole identified using monoclonal antibodies. Dev Biol 1987; 122:90-100.
7. Sperry RW. Chemoaffinity in the orderly growth of nerve fibre patterns and connection. Proc Natl Acad Sci USA 1963; 50:703-710.
8. Fujisawa H, Takagi S, Hirata T. Growth-associated expression of a membrane protein, neuropilin, in *Xenopus* optic nerve fibers. Dev Neurosci 1995; 17:343-349.
9. Ohta K, Mizutani A, Kawakami A et al. Plexin: A novel neuronal cell surface molecule that mediates cell adhesion via a homophilic binding mechanism in the presence of calcium ions. Neuron 1995; 14:1189-1199.
10. Ohta K, Takagi S, Asou H et al. Involvement of neuronal cell surface molecule B2 in the formation of retinal plexiform layers. Neuron 1992; 9:151-161.
11. Takagi S, Hirata T, Agata K et al. The A5 antigen, a candidate for the neuronal recognition molecule, has homologies to complement component and coagulation factors. Neuron 1991; 7:295-307.
12. Takagi S, Kasuya Y, Shimizu M et al. Expression of a cell adhesion molecule, neuropilin, in the developing chick nervous system. Dev Biol 1995; 170:207-222.
13. Kawakami A, Kitsukawa T, Takagi S et al. Developmentally regulated expression of a cell surface protein, neuropilin, in the mouse nervous system. J Neurobiol 1996; 29:1-17.
14. Chen H, Chédotal A, He Z et al. Neuropilin-2, a novel member of the neuropilin family, is a high affinity receptor for the semaphorins SemaE and SemaIV but not SemaIII. Neuron 1997; 19:547-559.
15. Maestrini E, Tamagnone L, Longati P et al. A family of transmembrane proteins with homology to the MET-hepatocyte growth factor receptor. Proc Natl Acad Sci USA 1996; 93:674-678.
16. Kameyama T, Murakami Y, Suto F et al. Identification of plexin family molecules in mice. Biochem Biophys Res Commun 1996; 226:396-402.
17. Kameyama T, Murakami Y, Suto F et al. Identification of a neuronal cell surface molecule, plexin, in mice. Biochem Biophys Res Commun 1996; 226:524-529.

18. Tamagnone L, Artigiani S, Chen H et al. Plexins are a large family of receptors for transmembrane, secreted, and GPI-anchored semaphorins in vertebrates. Cell 1999; 99:71-80.
19. Fujisawa H, Otsuki T, Takagi S et al. An aberrant retinal pathway and visual centers in *Xenopus* tadpoles share a common cell surface molecule, A5 antigen. Dev Biol 1989; 135:231-240.
20. Kitsukawa T, Shimono A, Kawakami A et al. Overexpression of a membrane protein, neuropilin, in chimeric mice causes anomalies in the cardiovascular system, nervous system and limbs. Development 1995; 121:4309-4318.
21. Fujisawa H, Kitsukawa T, Kawakami A et al. Roles of a neuronal cell surface molecule, neuropilin, in nerve fiber fasciculation and guidance. Cell Tiss Res 1997; 290:465-470.
22. Bagnard D, Lohrum M, Uziel D et al. Semaphorins act as attractive and repulsive guidance signals during the development of cortical projections. Development 1998; 125:5043-53.
23. Giger RJ, Urquhart ER, Gillespie SKH et al. Neuropilin-2 is a receptor for semaphorin IV: Insight into the structural basis of receptor function and specificity. Neuron 1998; 21:1079-1092.
24. Nakamura F, Tanaka M, Takahashi T et al. Neuropilin-1 extracellular domains mediate semaphorin D/III-induced growth cone collapse. Neuron 1998; 21:1093-1100.
25. Kitsukawa T, Shimizu M, Sanbo M et al. Neuropilin-semaphorin III/D-mediated chemorepulsive signals play a crucial role in peripheral nerve projection in mice. Neuron 1997; 19:995-1005.
26. Fujisawa H, Kitsukawa T. Receptors for collapsin/semaphorins. Current Opinion in Neurobiology 1998; 8:587-592.
27. Giger RJ, Cloutier JF, Sahay A et al. Neuropilin-2 is required in vivo for selective axon guidance responses to secreted semaphorins. Neuron 2000; 25:29-41.
28. Chen H, Bagri A, Zupicich JA et al. Neuropilin-2 regulates the development of selective cranial and sensory nerves and hippocampal mossy fiber projections. Neuron 2000; 25:43-56.
29. Campbell DS, Regan AG, Lopez JS et al. Semaphorin 3A elicits stage-dependent collapse, turning, and branching in *Xenopus* retinal growth cones. J Neurosci 2001; 21:8538-8547.
30. Takahashi T, Fournier A, Nakamura F et al. Plexin-neuropilin-1 complexes form functional semaphorin-3A receptors. Cell 1999; 99:59-69.
31. Rohm B, Ottemeyer A, Lohrum M et al. Plexin/neuropilin complexes mediate repulsion by the axonal guidance signal semaphorin 3A. Mech Dev 2000; 93:95-104.
32. Murakami Y, Suto F, Shimizu M, et al. Differential expression of plexin-A subfamily members in the mouse nervous system. Dev Dyn 2001; 220:246-258.
33. Soker S, Takashima S, Miao HQ et al. Neuropilin-1 is expressed by endothelial and tumor cells as an isoform-specific receptor for vascular endothelial growth factor. Cell 1998; 92:735-745.
34. Kawasaki T, Kitsukawa T, Bekku Y et al. A requirement for neuropilin-1 in embryonic vessel formation. Development 1999; 126:4885-4893.
35. Shimizu M, Murakami Y, Suto F et al. Determination of cell adhesion sites of neuropilin-1. J Cell Biol 2000; 148:1283-1294.
36. Satoda M, Takagi S, Ohta K et al. Differential expression of two cell surface proteins, neuropilin and plexin, in *Xenopus* olfactory axon subclasses. J Neurosci 1995; 15:942-955.
37. Kawasaki K, Bekku Y, Suto F et al. Requirement of neuropilin-1-mediated Sema3A signals in patterning of the sympathetic nervous system. Development 2002; 129:671-680.
38. Taniguchi M, Yuasa S, Fujisawa H et al. Disruption of semaphorin III/D gene causes severe abnormality in peripheral nerve projection. Neuron 1997; 19:519-530.
39. Castellani V, Chédotal A, Schachner M et al. Analysis of the L1-deficient mouse phenotype reveals cross-talk between Sema3A and L1 signaling pathways in axonal guidance. Neuron 2000; 27:237-249.

NEUROPILINS AS SEMAPHORIN RECEPTORS:

In vivo Functions in Neuronal Cell Migration and Axon Guidance

Anil Bagri[1] and Marc Tessier-Lavigne[2]

SUMMARY

After the initial discovery of neuropilin-1 as an epitope on axons recognized by a monoclonal antibody, neuropilins were rediscovered in the search for receptors mediating the repulsive actions of class 3 Semaphorins, notably Sema3A. Neuropilins are the ligand binding moieties in the class 3 Semaphorin receptor complexes, with the signaling moieties apparently provided by members of the plexin family. In their capacity as Semaphorin receptors, neuropilins have been shown to transduce repulsive guidance signals that direct a large variety of cell migration and axon guidance events. We summarize their demonstrated roles in driving axon fasciculation, channeling various axonal populations, inhibiting axonal branching, creating exclusion zones for axons, and providing directional guidance cues by being presented in gradients. In addition to their roles in repulsive axon guidance, evidence is accumulating that neuropilins also transduce some attractive guidance functions of Semaphorins.

INTRODUCTION

The previous Chapter described the initial identification of neuropilin-1 through a monoclonal antibody screen for epitopes with restricted distributions suggestive of roles in neural wiring, and the demonstration of an adhesive function for this protein (Fujisawa, Chapter 1 this book). In this Chapter, we describe how expression

[1]Department of Anatomy and of Biochemistry and Biophysics, Howard Hughes Medical Institute, University of California, San Francisco, CA 94143-0452 and [2]Department of Biological Sciences, Howard Hughes Medical Institute, Stanford University, Stanford, CA 94305-5020.

cloning approaches subsequently identified neuropilin-1 and -2 as components of receptors for class 3 semaphorins. We then turn to in vivo functions of neuropilins in nervous system wiring during development, all of which to date appear to reflect their roles as Semaphorin receptors.

IDENTIFICATION AND CHARACTERIZATION OF NEUROPILINS AS SEMAPHORIN RECEPTORS

Semaphorins are a Large Family of Axon Guidance Molecules

The semaphorins are a large family of transmembrane and secreted proteins initially identified as guidance molecules for developing axons. The first known member of this family, grasshopper Sema-1a, is a transmembrane protein identified as the epitope recognized by a monoclonal antibody that labeled particular axon fascicles (hence its original name, Fascilin IV), and which was implicated in restricting axon growth and preventing defasciculation at a particular boundary in the limb.[1] The identification of Sema-1a provided the starting point for the identification of a family of related molecules in insects and humans, including a key secreted family member in mammals, Sema3A.[2] Sema3A was also identified independently as a soluble protein from chicken brain capable of causing collapse of the growth cones of sensory axons (and initially called collapsin).[3] Interest in this family continued to grow with the findings that (i) there exists a large family of vertebrate Semaphorins,[4-6] (ii) *Drosophila* Sema II can function as a repellent in vivo,[7,8] and (iii) several secreted vertebrate Semaphorins function as potent repellents for various classes of axons.[4,9-14] At present, the metazoan Semaphorin family is divided into seven subfamilies (classes) defined by sequence and structural considerations, with over 20 known members in vertebrates (classes 3-7) (there are, additionally, several viral Semaphorins known) (Fig. 1).

Identification of Neuropilin-1 as a Sema3A Receptor

All known secreted Semaphorins in vertebrates fall into class 3, defined initially by Sema3A, which comprises six members (Sema3A - Sema3F, only five have been identified in mammals to date (Sema3D has been identified only in chicken and zebrafish)). Given the evidence that members of this class are potent repellents (discussed in more detail below), there was strong interest in identifying the receptor(s) that mediates their effects. Initial studies focused on identifying binding proteins for Sema3A, with several groups using the strategy of studying binding sites for a fusion of Sema3A with an easily detected epitope, either alkaline-phosphatase (Sema3A-AP fusion proteins), or the Fc portion of the human immunoglobulin molecule (Sema3A-Fc fusion proteins). These studies demonstrated the presence of Sema3A binding sites on sensory axons in vitro,[15,16] and on a variety of axonal tracts in vivo.[17-19] (Fig. 2).

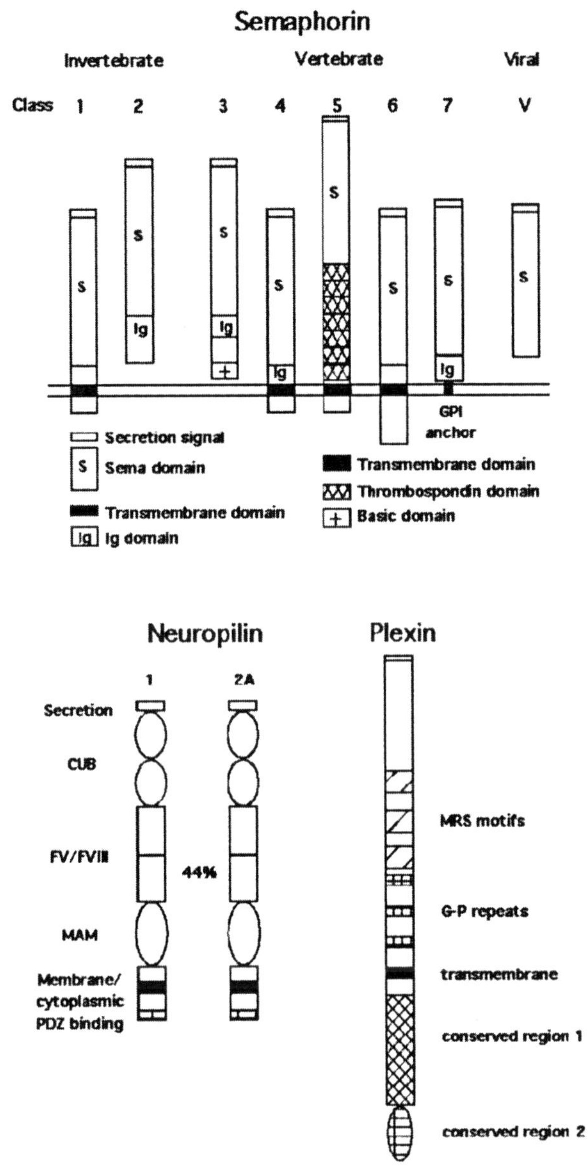

Figure 1. Schematic representation of the Semaphorin family and Semaphorin receptor families (reproduced from Nakamura et al., 2000).[54]
(Left): Semaphorins fall into seven subfamilies in animals (1-7), and are also found in certain viral genomes (V). All members of the family possess a Semaphorin (Sema) domain. Members of classes 1 and 4-6 are transmembrane, and those in class 7 are GPI anchored.
(Middle and Right): Semaphorin receptors include neuropilins (middle) and plexins (right; one of the 9 known mammalian plexins, Plexin-A1, is depicted). Their structure is discussed in detail in other Chapters.

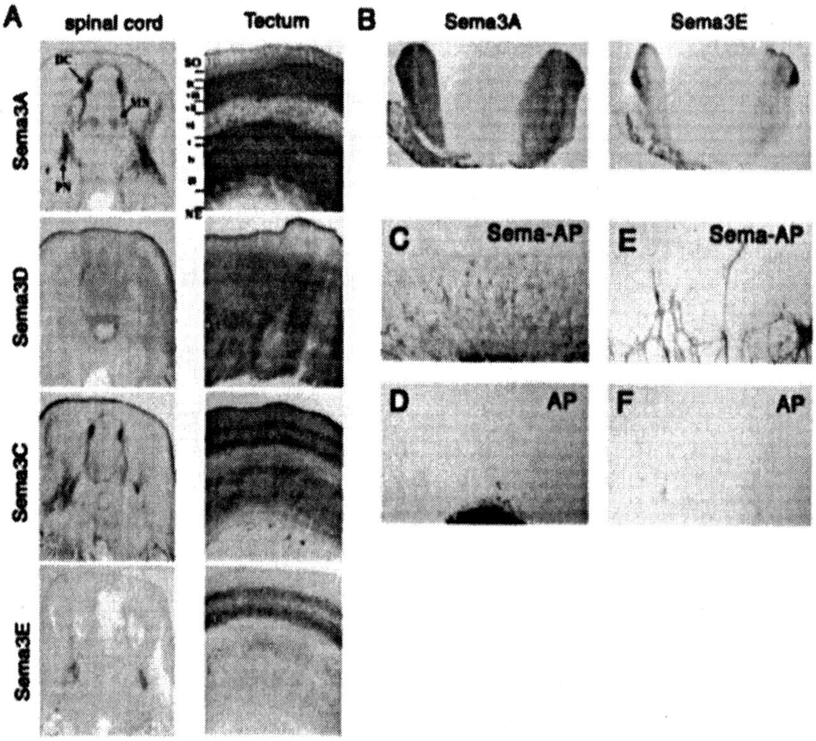

Figure 2. Binding of AP-tagged class 3 Semaphorins to tissue sections and cultured neurons (reproduced from Feiner et al, 1997; Kolodkin et al, 1997).[17,16]

(A) AP-tagged Sema3A, -3D, -3C and -3E label different neural structures on transverse sections of stage 27 chick spinal cord and E10 tectum. Similar patterns of labeling are observed for AP-Sema3A and AP-Sema3C (the latter labeling more weakly than the former). AP-Sema3E has a very restricted binding pattern. Tectal layers are indicated by Roman numerals. Other abbreviations: DC, dorsal columns; MN, motoneurons; PN, peripheral nerves; SO, stratum opticum.

(B) A tract in anterior diencephalon that is bound by AP-Sema3E, but is not stained by AP-Sema3A.

(C-F) DRG explants obtained from E14 rat embryos were grown in tissue culture for two days in the presence of NGF, then processed for in situ binding by Sema–AP (C and E), or by a control construct (secreted alkaline phosphatase) (D and F). Note that Sema–AP binding activity is detected on axons and growth cones of DRG neurons. Scale bar = 100 μm in (C), (D), 25 μm in (E) and (F).

The finding of selective binding of Sema3A-AP to particular axonal populations prompted two groups to attempt to identify the relevant binding protein(s) through expression cloning in COS cells.[15,16] Both made cDNA libraries from appropriately staged embryonic rat sensory ganglia in a COS cell expression vector, divided the library into pools of 750 -2000 plasmids, transfected these pools into COS cells, and probed the transfected cells for the presence of Sema3A-AP binding sites. Through a screen of 140 and 70 pools, respectively, the groups each identified a positive pool, and subdivided it through several rounds until each group identified a single plasmid which, when transfected into COS cells, created Sema3A-AP binding sites on the cells.[15,16]

In both cases, the active plasmid was found to encode rat neuropilin-1,[15,16] providing the first evidence that neuropilins are class 3 Semaphorin receptors. The binding coefficient (Kd) for the Sema3A - neuropilin-1 interaction on COS cells was found to be ~0.3 nM, sufficient to account quantitatively for the binding observed on isolated sensory neurons. Two types of functional data supported a necessary role for neuropilin-1 in mediating the repulsive actions of Sema3A. First, antibodies to neuropilin-1 were found to block the ability of Sema3A both to cause collapse of growth cones of sensory axons and to repel these axons in a three dimensional collagen gel[15,16] (Fig. 3). Second, the analysis of Sema3A and neuropilin-1 knock-out mice demonstrated striking similarities in axon guidance phenotypes in the mutant embryos of both genotypes (discussed in more detail below).[20,21] Finally, tying the knock-out mice to the in vitro assays, it was shown that sensory neurons isolated from neuropilin-1 mutant embryos fail to respond to Sema3A,[21] consistent with the evidence from function-blocking antibodies that neuropilin-1 is a necessary receptor for Sema3A in axon guidance.

Differential Actions of Class 3 Semaphorins Mediated by Neuropilin-1 and Neuropilin-2

Initial studies of class 3 Semaphorins demonstrated that different members of this class have differential effects on different classes of neurons. For instance, Sema3A causes collapse and/or repels both embryonic sensory and sympathetic growth cones[3,4,9] whereas Sema3C and Sema3F have such effects principally on sympathetic but not sensory growth cones.[22,23] Furthermore, in studies using AP fusion proteins, the existence of different binding sites on tissue sections for different class 3 Semaphorins was revealed.[17] Together, these studies suggested that the differential responses of different neurons to class 3 Semaphorins might be mediated by different receptors.

A candidate for a receptor that might mediate differential responses to Semaphorins was provided by the identification, through sequence homology, of a second member of the neuropilin family, neuropilin-2.[16,24] Neuropilin-2 was found to possess several isoforms arising from alternative splicing, that can result in either of two intracellular domains (a and b isoforms), and in the insertion of short stretches encoded in small exons in the extracellular region near the transmembrane domain, although the functional consequences of this splicing is unknown. Importantly, differential binding of Sema3A and Sema3F was observed, with Sema3A binding with high affinity and preferentially to neuropilin-1 (at least 30 times more avidly than to neuropilin-2), and Sema3F binding with high affinity and preferentially to neuropilin-2 (at least 10 times more avidly than to neuropilin-1).[24] This differential binding appeared to explain the specificity of action of the two Semaphorins. Thus, neuropilin-1 is expressed by both sensory and sympathetic axons, which both respond to Sema3A, whereas neuropilin-2 is expressed only by sympathetic, not sensory axons, and its high affinity ligand Sema3F similarly repels only sympathetic, not sensory axons.

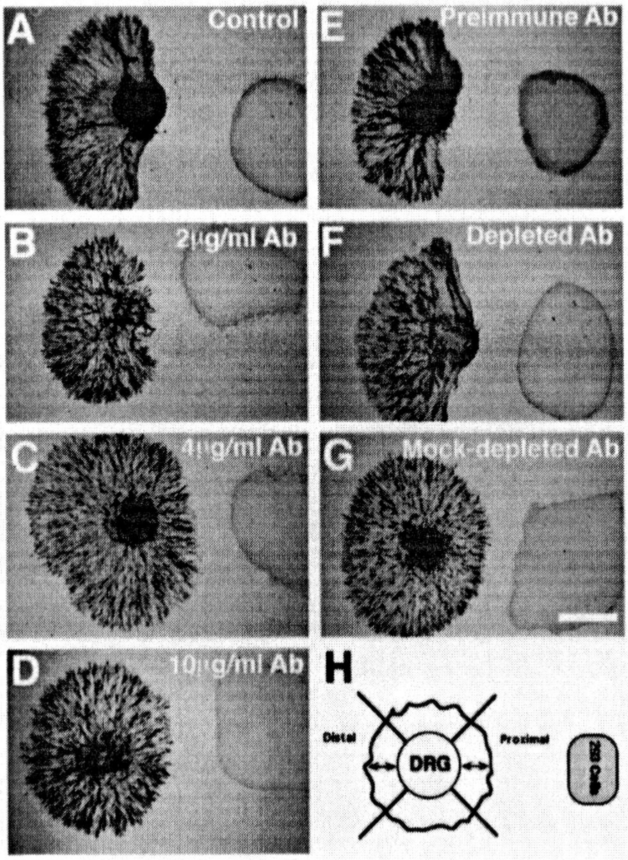

Figure 3. Functional requirement of neuropilin-1 for Sema3A-evoked repulsion of NGF-responsive DRG axons (reproduced from He and Tessier-Lavigne (1997)).[15]
E14 rat DRG explants were cultured in collagen gels with 25 ng/ml NGF to elicit outgrowth of Sema III–responsive axons (Messersmith et al 1995). Explants were cocultured with aggregates of 293-EBNA cells secreting Sema3A–AP protein (right in each panel) in the presence of 0 μg/ml (A), 2 μg/ml (B), 4 μg/ml (C), or 10 μg/ml (D) of anti-neuropilin IgG, 10 μg/ml of preimmune IgG (E), or 10 μg/ml of depleted (F) or mock-depleted (G) anti-neuropilin IgG for 40 hr. The explants were then fixed and visualized by wholemount immunostaining with the anti-neurofilament antibody NF-M. DRG neurites proximal to, but not distal to, the Sema3A–AP–secreting cells were repelled in the absence of anti-neuropilin antibody, an effect that was blocked in a dose-dependent fashion by addition of the antibody. (H) shows the procedure used to quantify the reponse. Scale bar: 350 μm. A color version of this figure is included in the insert that precedes Chapter 1.

Together, these results suggested a model[24] (Fig. 4) in which neuropilin-1 is a high affinity receptor for Sema3A and neuropilin-2 a high affinity receptor for Sema3F, with the differential actions of these two Semaphorins on different neuronal populations dictated by the complement of neuropilins expressed by the neurons. This model was confirmed by studies using function-blocking antibodies

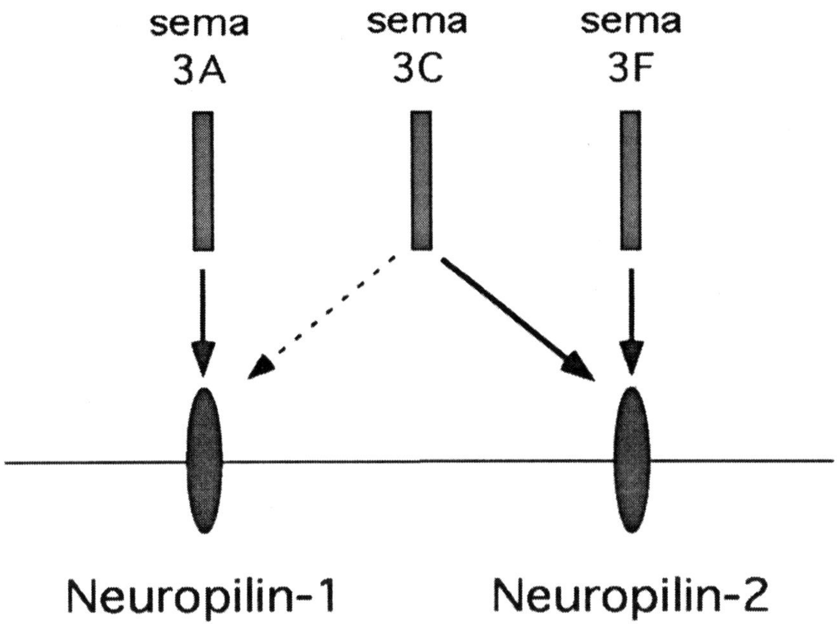

Figure 4. Schematic representation of receptor specificity of different class 3 semaphorins. Sema3A signaling is mediated via neuropilin-1, whereas sema3F signaling is mediated via neuropilin-2. Sema3C mediated signaling requires both neuropilins, but requires neuropilin-2 to a greater extent than neuropilin-1 (hence the dotted arrow). A color version of this figure is included in the insert that precedes Chapter 1.

which showed that neuropilin-1 but not neuropilin-2 is required for repulsive actions of Sema3A on sympathetic axons,[25] whereas neuropilin-2 but not neuropilin-1 is required for repulsive actions of Sema3F on those axons.[25,26] The actions of the function-blocking antibodies were further confirmed using sympathetic[27,28] and hippocampal neurons[28] isolated from neuropilin-2 knock-out mice, which also lost their responsiveness to Sema3F.

A slightly more complicated version of this model has been invoked to account for the actions of Sema3C, which binds both neuropilin-1 and neuropilin-2 equally.[24] It is thought that Sema3C absolutely requires neuropilin-2 for its function but requires neuropilin-1 to a lesser extent. This conclusion is based on the observation that Sema3C does not repel sensory axons (which express only neuropilin-1) but it does repel sympathetic axons (which express both), and antibodies to neuropilin-1 decrease Sema3C-induced repulsion of sympathetic axons by ~80% but do not block it entirely. The model that is suggested by these observations is that a receptor comprising only neuropilin-2 not neuropilin-1 can transduce the Sema3C signal to some extent, but efficient transduction of the signal requires both neuropilins (Fig. 4). Sema3B also appears to bind both neuropilins about equally, and may function in a similar way to Sema3C.[29] The fact that class 3 Semaphorins appear to function as cross-linked dimers,[19,30,31] suggests that co-receptors of neuropilin-1 and neuropilin-2 may to some extent be induced by Sema3B and Sema3C themselves, although co-immunoprecipitation experiments in transfected cells show that the two neuropilins can also associate with one another in a ligand-independent fashion.[23,26,29]

In addition to the loss-of-function experiments using antibodies, gain-of-function experiments in which neuropilins were delivered to different neuronal populations using recombinant herpes simplex viruses provided further support for the specificity model.[29] Thus, expression of neuropilin-2 in chick sensory neurons, which normally only express neuropilin-1 and only respond to Sema3A, made these cells responsive to Sema3B and Sema3C. Inversely, expressing neuropilin-1 in chick retinal ganglion cells, which normally do not express neuropilins and do not respond to Sema3A, made these cells responsive to Sema3A.

Finally, structure-function studies showed that the specificity of collapsing actions of Sema3A and Sema3C on sensory and sympathetic growth cones (with Sema3A affecting both and Sema3C only sympathetic neurons), is conferred by a 70 amino acid stretch within the Semaphorin domains of both proteins.[22] An elucidation of the structural aspects of Semaphorins and neuropilins that dictate the specificity of action of the different ligands awaits future studies.

Neuropilins are Binding Moieties in a Receptor Complex with Plexins

When neuropilins were initially identified as class 3 Semaphorin receptors, the short length of the cytoplasmic tail of these proteins prompted speculation that they might only function as ligand-binding moieties in receptor complexes comprising additional proteins as signaling moieties. This idea was strengthened by the finding that the cytoplasmic domain of neuropilin-1 is apparently dispensable. This was shown using a mutated form of neuropilin-1 in which the transmembrane and cytoplasmic domain were replaced by a glycosyl-phosphatidylinositol linkage sequence. This protein was delivered to chick retinal ganglion cells (which do not normally express neuropilin-1) using a viral vector, and found to be expressed on the surface of the cells (as expected) and to confer Sema3A responsiveness to these neurons, despite the absence of the neuropilin cytoplasmic domain.[55]

Subsequent studies identified plexins as signaling proteins that complex with neuropilins to mediate the repulsive actions of class 3 Semaphorins.[32-36] This function of plexins is reviewed in detail in Chapter 5 (Püschel AW, this book), and so is not discussed in any more detail here, nor is the possibility that other molecules such as the adhesion molecule L1 might be part of Semaphorin receptor complexes, a possibility reviewed in Chapters 6 and 7 (Neufeld G et al; Castellani V, this book).

IN VIVO FUNCTIONS OF NEUROPILINS IN NERVOUS SYSTEM WIRING DURING DEVELOPMENT

A combination of in vitro studies, embryological studies in chicken embryos, and genetic analysis in mice has suggested important roles for neuropilins as receptors mediating repulsive actions of class 3 Semaphorins to direct various aspects of nervous system wiring. To facilitate a review of the literature, we break down the suggested functions of neuropilins into three categories: regulation of axon fasciculation, regulation of axon guidance and cell migration through creation of exclusion zones, and directional guidance of axons and dendrites based on detection of Semaphorin gradients. In some cases, we discuss what is known about the actions of particular Semaphorins even if the involvement of neuropilins is only inferred, not demonstrated.

Regulation of Axon Fasciculation, Channeling and Branching

As first proposed in a reinterpretation of experiments on ephrins,[37] the presence of a repellent factor in the environment of growing axons can help to drive axon fasciculation by making axons prefer to grow on the surface of other axons rather than the surface of cells in the environment. It is interesting in this context that the first functional perturbation of a Semaphorin in vivo, the transmembrane grasshopper Sema-1a, resulted in defasciculation and sprouting of sensory axons that normally grow in contact with the Semaphorin.[1] Perturbation of Semaphorin function results in defasciculation. In both Sema3A and neuropilin-1 knock-out mice, profound defasciculation of trigeminal sensory axons, as well as other cranial and spinal sensory axons, was reported,[20,21] consistent with Sema3A in the environment of these axons driving the axons to fasciculate with one another (Fig. 5). It is not known whether the fasciculation is driven by Sema3A present uniformly in the environment, or whether some graded distribution contributes to the fasciculation; in particular, it has been proposed that graded distribution of repellent molecules flanking sensory axons might contribute to channeling the axons together through a process of "surround-repulsion",[38] which could in principle involve Sema3A. Similarly, in neuropilin-2 knock-out mice, defasciculation of cranial nerve III (oculomotor) axons and axons of the ophthalmic branch of cranial nerve V (trigeminal) was observed.[27,28] More recently, severe defasciculation of vomeronasal sensory axons en route to the accessory olfactory bulb was also observed in neuropilin-2 knock-out mice.[39] Sema3A was also found to inhibit the branching of cortical axons growing on two dimensional substrates,[12] which, though a slightly different cell biological phenomenon, is likely another manifestation of the ability of Semaphorins, acting via neuropilins, to make substrates less favorable for growth and to drive fasciculation. Finally, in an analysis of encounters between thalamic and cortical axons, Sema3A was shown not just to drive fasciculation, but also to potentiate the effects of other factors that appear to drive selective fasciculation of these axons with others of like origin (thalamic with thalamic and cortical with cortical).[40]

Generation of Exclusion Zones

The simplest guidance role for a putative repellent molecule is to create an exclusion zone that bars entry of responsive axons into an inappropriate region. Such a role has been proposed for class 3 Semaphorins, acting via neuropilins, for

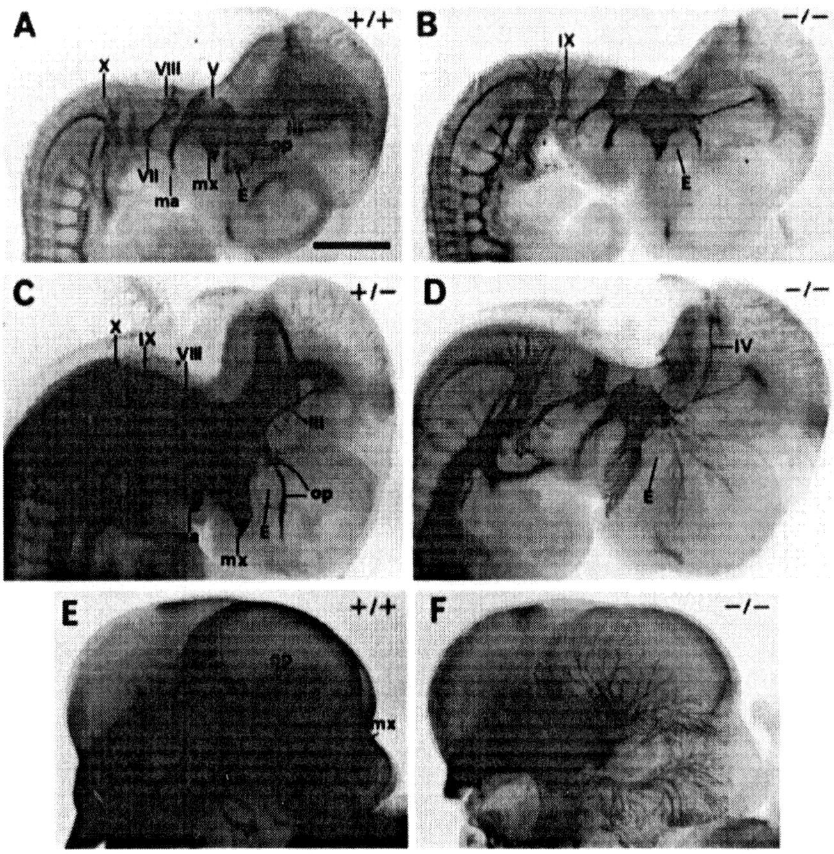

Figure 5. Defects in projections of cranial nerves in neuropilin-1 mutant embryos (reproduced from Kitsukawa et al, 1997).[20]
Panels show whole-mount immunostaining with anti-neurofilament monoclonal antibody 2H3 of wild-type (+/+), heterozygous (+/-), and homozygous mutant (-/-) embryos at E9.5 (A and B), E10.5 (C and D), and E12.5 (E and F), to reveal defects in cranial nerves. III, oculomotor nerve; IV, trochlear nerve; V, trigeminal nerve; VII, facial nerve; VIII, vestibulocochlear nerve; IX, glossopharyngeal nerve; X, vagus nerve; op, ophthalmic nerve; mx, maxillary nerve; ma, mandibular nerve, E; eye. Scale bar, 1 mm. A color version of this figure is included in the insert that precedes Chapter 1.

many axonal populations. As we review, this function has been confirmed in many—but not all—cases.

Exclusion Zones for Sensory Axon Collaterals in the Gray Matter

One of the first roles proposed for Sema3A was to generate an exclusion zone in the spinal cord for a subset of sensory axon collaterals. Sensory axons in the dorsal root ganglia extend axons to the dorsal edge of the spinal cord (the dorsal root entry zone) and send axons alongside the dorsal spinal cord in the dorsal funiculus for several days before sprouting collaterals into the spinal cord gray matter. The collaterals of different functional subclasses of sensory neurons have different laminar termination sites. Thus, large-diameter proprioceptive neurons, which are responsive to Neurotrophin-3 and express the NT-3 receptor trkC, terminate on motoneurons in the ventral spinal cord, whereas small-diameter NGF-responsive sensory neurons that express trkA terminate in the dorsal spinal cord. Sema3A transcripts were found to be expressed in the ventral spinal cord at the time that this patterning of terminations is occurring, and differential responses of the two classes of neurons were demonstrated: the NGF-responsive sensory neurons very profoundly repelled by Sema3A in vitro, whereas the NT-3 responsive sensory ones were found to be much less responsive.[9] The repulsive action of Sema3A is mediated by neuropilin-1, which is expressed by the small diameter but not the large diameter collaterals.[15,16] [41] These results suggested that Sema3A, acting via neuropilin-1, might normally function to create an exclusion zone selectively for the NGF-responsive sensory axons, preventing them but not the NT-3 responsive axons from invading the ventral spinal cord;[9,42] the selectivity would be conferred by differential expression of neuropilin receptors by the two classes of neurons. Analysis of a Sema3A knock-out mouse demonstrated, as predicted, that some small diameter sensory collaterals (defined by expression of CGRP) projected abnormally ventrally,[43] but later analysis of an independently derived Sema3A knock-out mouse failed to demonstrate extensive errors of projection of sensory collaterals visualized with DiI.[21] Thus, if Sema3A functions to exclude small diameter collaterals, it must be just one of several redundant cues. Further, experiments in chick embryos involving misexpression of Sema3A, suggested that Sema3A does not diffuse far,[41] so that a role in repulsion of small diameter sensory axons may involve principally those that inappropriately overshoot their normal termination site.

Exclusion Zones for Peripheral Branches of Sensory Axons

A more robust role for Sema3A in creating an exclusion zone has been documented in the case of the peripheral branches of trigeminal sensory axons, since in Sema3A or neuropilin-1 knockout mice, trigeminal sensory axons projecting in the ophthalmic branch of the trigeminal ganglion overshoot their termination site, which is normally a site of expression of Sema3A.[20,21]

Creating a "Waiting Period" for Olfactory Axon Invasion of the Olfactory Bulbs

Another clear demonstration of a role for Sema3A in creating an exclusion zone was obtained in the chick olfactory system.[44] Primary olfactory neurons connect to the olfactory bulb. During development, their axons extend and reach the bulb several days before the target matures, and they then experience a "waiting period", accumulating and staying outside the target, and only entering several days later when the target matures (Fig. 6). During this waiting period, Sema3A transcripts are expressed in the target, and the olfactory axons express neuropilin-1 and are responsive to Sema3A in vitro. The function of Sema3A in this context could not be analyzed in Sema3A or neuropilin-1 knock-out mice, since they die too early, so it was instead studied in chick embryos. Misexpression of a dominant-negative form of neuropilin-1 in these neurons by electroporation allowed many of their axons to enter the olfactory bulbs prematurely[44] (Fig. 6), providing evidence that Sema3A is responsible, at least in part, for preventing the axons from entering the target during the waiting period.

Excluding Commissural Axons from the Midline and Gray Matter

Commissural axons in the spinal cord extend through the spinal cord gray matter from their dorsally located cell bodies to floor plate cells at the ventral midline, then cross the midline, and, after crossing, make a sharp turn to enter the ventral funiculus, thereby exiting the gray matter. Two class 3 Semaphorins, Sema3B and Sema3F, acting via a neuropilin-2 containing receptor, have been implicated in helping expel the axons from the midline and the gray matter.[45] Sema3B is expressed by floor plate cells, whereas Sema3F is expressed broadly in the marginal zone of the spinal cord gray matter, excluding the floor plate. Commissural neurons express neuropilin-2 mRNA, but they are insensitive to Sema3B and Sema3F prior to reaching the floor plate, only becoming responsive (through an unknown mechanism) after crossing the midline (Fig. 7). The ability of Sema3B and Sema3F to repel these axons after they cross could contribute to expelling them from the midline (Sema3B) and the spinal cord gray matter (Sema3F) after crossing, and help push them into the ventral funiculus. This model was supported by analysis of a neuropilin-2 knock-out mouse, in which stalling of commissural axons at the midline at high penetrance was observed[45] (Fig. 7), consistent with a normal role for Sema3B in expelling the axons from the midline.

Excluding Ipsilaterally-Projecting Axons from the Ventral Midline Region

Sema3A has also been implicated in preventing dorsal tectal neurons that form the tectobulbar tract from sending axons too far ventral.[46] These axons normally project circumferentially along a ventral trajectory, but then turn to project longitudinally (and caudally) without crossing the medial longitudinal fasciculus (MLF), which courses alongside the floor plate. These axons can be repelled by Sema3A but not Sema3B or 3C, and are repelled by tissue containing MLF neurons, which appear to be a source of Sema3A. In Sema3A knock-out mice, tectobulbar axons cross the MLF rather than turning caudally, consistent with Sema3A creating an exclusion zone that forces them to switch from a circumferential to a longitudinal trajectory.[46]

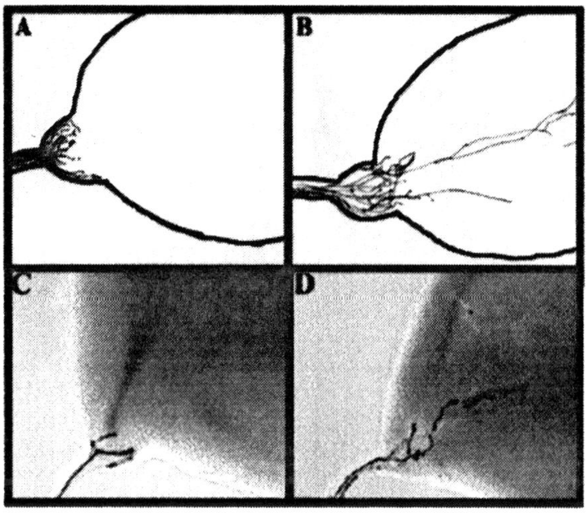

Figure 6. Neuropilin enforces a waiting period for olfactory axons at the olfactory bulb in the developing chick embryo (reproduced from Renzi et al, 2000).[44]
Electroporation was used to introduce a control alkaline phosphatase construct (A, C), or a dominant negative neuropilin-1 construct together with the alkaline phosphatase construct (B, D) into embryonic chick olfactory neurons. Labeled axons are visualized by alkaline phosphatase histochemistry in whole-mounted brains. In each of (A, B), trajectories of olfactory axons from four embryos are shown as a composite drawing; raw data for one embryo are shown in (C, D). Olfactory axons expressing the dominant-negative neuropilin-1 construct (B, D) enter the telencephalon prematurely.

Sorting Migrating Interneurons

Sema3A and Sema3F appear to collaborate to create an exclusion zone that helps sort migrating cortical interneurons from striatal interneurons.[47] Transcripts for both factors are expressed by the striatal primordium, and both neuropilin receptors are expressed selectively by interneurons targeting the cortex but not the striatum. Those cortical interneurons are, as expected, repelled by striatal tissue as well as by cells expressing the two Semaphorins. Loss of neuropilin function (achieved using a neuropilin-2 knock-out mouse, and dominant-negative neuropilin constructs) increases the number of interneurons that migrate into the striatum, consistent with a role for the two Semaphorins in preventing cortical interneurons from targeting the striatum.[47]

Directional Guidance Based on Detection of Semaphorin Gradients

In addition to regulating fasciculation, channeling axons, and creating exclusion zones, class 3 Semaphorins, acting via neuropilins, have also been proposed to guide axons by being presented in gradients that impart directionality on axons and cells.

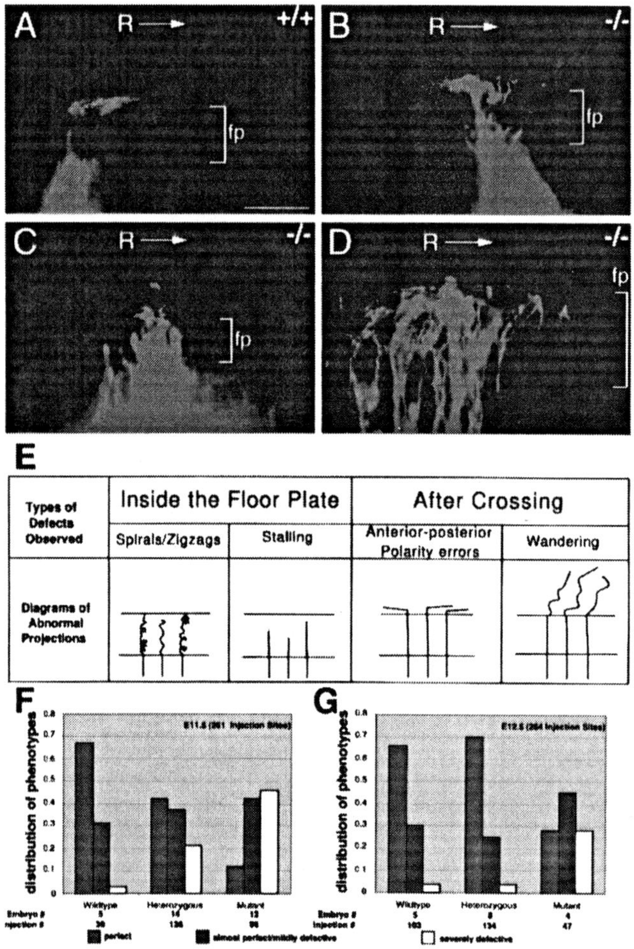

Figure 7. Neuropilin-2 Is Required for Normal Midline Commissural Axon Pathfinding In Vivo in the Developing Spinal cord (reproduced from Zou et al, 2000).[45]
(A–D) Visualization of commissural axon behavior at the floor plate (fp) in a wild-type E11.5 mouse embryo (A) and in three homozygous mutant neuropilin-2 E11.5 mouse embryos (B-D). Commissural axons are visualized following DiI injection in the dorsal spinal cord (off the bottom in each panel) in the "open book" configuration. Rostral (R) is to the right in each panel (indicated by arrow). In wild-type (A), commissural axons cross and turn rostrally in a very stereotyped fashion. A first example of pathfinding in a mutant embryo (B) shows randomization of the anterior–posterior projection patterns of commissural axons after exiting the floor plate, wavy axons and stalling growth cones inside the floor plate (note that the ''waviness'' starts approximately at the floor plate). A second example (C) shows commissural axons that are overshooting and wandering into the contralateral ventral spinal cord region after floor plate crossing. A third example (D) shows spiraling and wavy trajectories inside the floor plate (note again that the waviness is seen inside the floor plate, not before the floor plate). Scale bar: (A-C), 100 μm; (D), 66.7 μm. (E) Summary of commissural misrouting phenotypes in neuropilin-2 mutant mice. (F, G) Penetrance of defects in E11.5 and E12.5 embryos. A color version of this figure is included in the insert that precedes Chapter 1.

The strongest case for guidance through a gradient is provided by the analysis of Sema3A effects on cortical pyramidal neurons. These neurons express neuropilin-1, and their axons are repelled by Sema3A.[12,48] Normally, these axons project away from the pia towards the ventricular surface. The marginal zone underneath the pia was shown to possess a repulsive activity for these axons through experiments in which labeled cortical neurons were seeded on cortical slices.[48] They were found to extend axons away from a nearby marginal zone, but to extend axons in random directions if they were positioned on deep cortical layers at a distance from the marginal zone. This repellent activity was proposed to be mediated by Sema3A since it was abolished by antibodies to neuropilin-1, and since Sema3A transcripts were found in the cortex. In Sema3A knock-out mice, the normal polarity of cortical pyramidal axons was found to be disrupted,[48] consistent with the model that a gradient of Sema3A emanating from the pial marginal zone repels these axons towards the ventricular surface (Fig. 8). Interestingly, the ability of Semaphorin gradients to direct axon growth is not very sensitive to the precise shape and concentration of the factor, ensuring that the guidance is robust.[49]

A function for Sema3A in repelling axons from behind has also been suggested in the case of the migrations of trochlear and brachial motor axons away from the ventral midline in the hindbrain.[10] Sema3A has also been proposed to repel a population of glial progenitor cells away from the optic chiasm into the optic nerve, based on the presence of Sema3A transcripts in the chiasm, and the responsiveness of these cells to Sema3A.[50] However, functional perturbations in vivo have not yet been performed to test these hypotheses.

Attractive Functions of Semaphorins

We have, so far, concentrated on repulsive and inhibitory actions of class 3 Semaphorins, mediated by neuropilin-containing receptors. There is, however, mounting evidence that some class 3 Semaphorins may also have positive—attractive—functions. Evidence for such functions came from studies of cortical neurons, whose axons, as discussed above, are repelled by Sema3A, migrating away from the source. It was found, in contrast, that the cortical axons will orient up a gradient of the class 3 Semaphorin Sema3C; this attractive response was only seen when a gradient of Sema3C was formed.[12,49] Similarly, an attractive effect of Sema3B was observe on axons of olfactory bulb neurons (which, as described above, are repelled by Sema3F).[51] These intriguing observations naturally raise the question of whether neuropilin receptors contribute to these responses, an issue that has not so far been addressed.

Importantly, individual neurons can respond to a given class 3 Semaphorin with either repulsion or attraction depending on the status of cGMP signaling in the growth cone.[52] This was shown for Xenopus spinal neurons in culture, which are normally repelled by Sema3A via a neuropilin-containing receptor. Manipulations that increase cGMP signaling in the growth cone (such as addition of a membrane permeable analogue, dibutyrl cGMP), convert this repulsion to attraction, an effect still

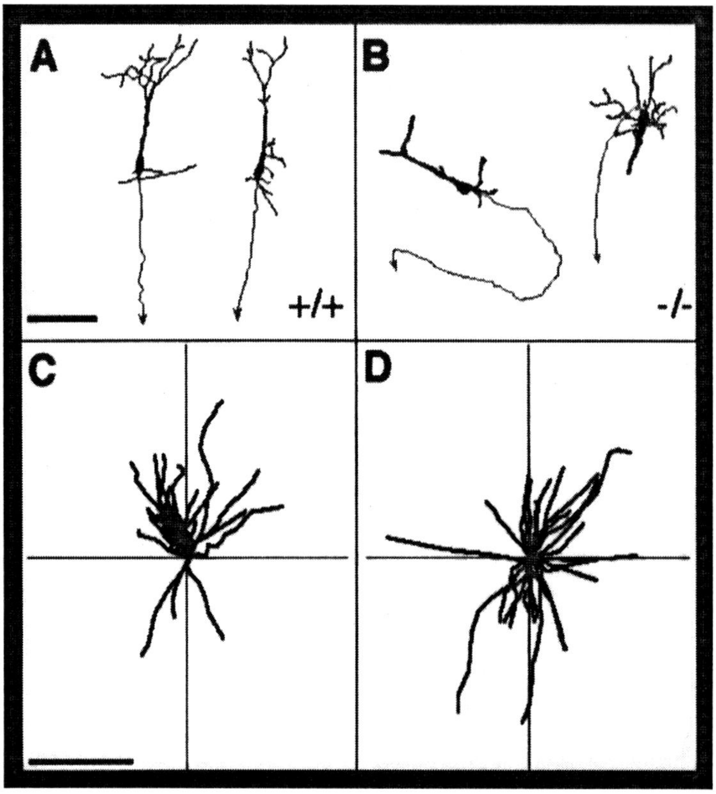

Figure 8. Repulsion of axons and attraction of dendrites of cortical pyramidal neurons by Sema3A, activating neuropilin-1 (reproduced from Polleux et al, 1998; 2000).[48,53]
Top panels: Evidence that endogenous Sema3A contributes to cortical axon guidance. Examples of the morphologies of cortical neurons in layers V-VI in wild-type (left) and Sema3A knock-out (right) mice labeled by white matter DiA injections. The pia is to the top; axons are indicated in red. Scale bar, 100 μm. Bottom panels: Evidence that neuropilin-1 is required for correct apical dendrite orientation. Shown are apical dendrite orientation plots of E15 neurons from mice expressing GFP under the beta-actin promoter (and which express GFP) cultured on postnatal cortical slices in the absence (left) or presence (right) of a function-blocking antibody to neuropilin-1. The pia is to the top. Note that normally the dendrites are oriented towards the pia, but the anti-neuropilin-1 antibody interferes with this directionality. Scale bar, 200 μm. A color version of this figure is included in the insert that precedes Chapter 1.

requiring neuropilin-1. Similarly, db-cGMP could partially block the collapsing activity of Sema3A on rat sensory growth cones.[52]

A physiological context for this ability to convert Sema3A signaling was provided by studies of the apical dendrites of cortical pyramidal neurons, which take a trajectory that is essentially opposite to that of the axons of these neurons, i.e., towards the pial surface and away from the ventricular zone. These dendrites were found to possess high concentrations of guanylate cyclase, the enzyme catalyzing the synthesis of cGMP, and they were found to be attracted by Sema3A rather than repelled.[53] Thus, a single cue is proposed to have one effect on one of the cell's processes (repulsion of the axon) and the opposite effect on another of the cell's processes (attraction of the dendrite), presumably dictated by the level of cGMP in the process (Fig. 8).

CONCLUSION

In this Chapter we have reviewed the known and proposed functions of neuropilins in neuronal cell migration and axon guidance. To date, all these functions relate to the known role of neuropilins as Semaphorin receptors. Although the best-characterized functions are in cell and axonal repulsion, the final section has emphasized the possibility that attractive responses of neurons and axons might also be mediated by neuropilins. Furthermore, the possibility remains that some functions of neuropilins are independent of Semaphorins. Future analysis of neuropilin mutants animals should help shed light on these possibilities.

REFERENCES

1. Kolodkin AL, Matthes DJ, O'Connor TP et al. Fasciclin IV: Sequence, expression, and function during growth cone guidance in the grasshopper embryo. Neuron 1992; 9:831-845.
2. Kolodkin AL, Matthes DJ, Goodman CS. The semaphorin genes encode a family of transmembrane and secreted growth cone guidance molecules. Cell 1993; 75:1389-1199.
3. Luo Y, Raible D, Raper JA. Collapsin: A protein in brain that induces the collapse and paralysis of neuronal growth cones. Cell 1993; 75:217-227.
4. Puschel AW, Adams RH, Betz H. Murine semaphorin D/collapsin is a member of a diverse gene family and creates domains inhibitory for axonal extension. Neuron 1995; 14:941-948.
5. Luo Y, Shepherd I, Li J et al. A family of molecules related to collapsin in the embryonic chick nervous system. Neuron 1995; 14:1131-1140.
6. Adams RH, Betz H, Puschel AW. A novel class of murine semaphorins with homology to thrombospondin is differentially expressed during early embryogenesis. Mech Dev 1996; 57:33-45.
7. Matthes DJ, Sink H, Kolodkin AL et al. Semaphorin II can function as a selective inhibitor of specific synaptic arborizations. Cell 1995; 81:631-639.
8. Winberg ML, Noordermeer JN, Tamagnone L et al. Plexin A is a neuronal semaphorin receptor that controls axon guidance. Cell 1998; 95:903-916.
9. Messersmith EK, Leonardo ED, Shatz CJ et al. Semaphorin III can function as a selective chemorepellent to pattern sensory projections in the spinal cord. Neuron 1995; 14:949-959.
10. Varela-Echavarria A, Tucker A, Puschel AW et al. Motor axon subpopulations respond differentially to the chemorepellents netrin-1 and semaphorin D. Neuron 1997; 18:193-207.
11. Kobayashi H, Koppel AM, Luo Y et al. A role for collapsin-1 in olfactory and cranial sensory axon guidance. J Neurosci 1997; 17:8339-8352.
12. Bagnard D, Lohrum M, Uziel D et al. Semaphorins act as attractive and repulsive guidance signals during the development of cortical projections. Development 1998; 125:5043-5053.
13. Chedotal A, Del Rio JA, Ruiz M et al. Semaphorins III and IV repel hippocampal axons via two distinct receptors. Development 1998; 125:4313-4323.
14. Rabacchi SA, Solowska JM, Kruk B et al. Collapsin-1/semaphorin-III/D is regulated developmentally in Purkinje cells and collapses pontocerebellar mossy fiber neuronal growth cones. J Neurosci 1999; 19:4437-4448.
15. He Z, Tessier-Lavigne M. Neuropilin is a receptor for the axonal chemorepellent Semaphorin III. Cell 1997; 90:739-751.

16. Kolodkin AL, Levengood DV, Rowe EG et al. Neuropilin is a semaphorin III receptor. Cell 1997; 90:753-762.
17. Feiner L, Koppel AM, Kobayashi H et al. Secreted chick semaphorins bind recombinant neuropilin with similar affinities but bind different subsets of neurons in situ. Neuron 1997; 19:539-545.
18. Takahashi T, Nakamura F, Strittmatter SM. Neuronal and non-neuronal collapsin-1 binding sites in developing chick are distinct from other semaphorin binding sites. J Neurosci 1997; 17:9183-9193.
19. Eickholt BJ, Morrow R, Walsh FS et al. Structural features of collapsin required for biological activity and distribution of binding sites in the developing chick. Mol Cell Neurosci 1997; 9:358-371.
20. Kitsukawa T, Shimizu M, Sanbo M et al. Neuropilin-semaphorin III/D-mediated chemorepulsive signals play a crucial role in peripheral nerve projection in mice. Neuron 1997; 19:995-1005.
21. Taniguchi M, Yuasa S, Fujisawa H et al. Disruption of semaphorin III/D gene causes severe abnormality in peripheral nerve projection. Neuron 1997; 19:519-530.
22. Koppel AM, Feiner L, Kobayashi H et al. A 70 amino acid region within the semaphorin domain activates specific cellular response of semaphorin family members. Neuron 1997; 19:531-537.
23. Giger RJ, Pasterkamp RJ, Holtmaat AJ et al. Semaphorin III: role in neuronal development and structural plasticity. Prog Brain Res 1998; 117:133-149.
24. Chen H, Chedotal A, He Z et al. Neuropilin-2, a novel member of the neuropilin family, is a high affinity receptor for the semaphorins Sema E and Sema IV but not Sema III. Neuron 1997; 19:547-559.
25. Giger RJ, Urquhart ER, Gillespie SK et al. Neuropilin-2 is a receptor for semaphorin IV: Insight into the structural basis of receptor function and specificity. Neuron 1998; 21:1079-1092.
26. Chen H, He Z, Bagri A et al. Semaphorin-neuropilin interactions underlying sympathetic axon responses to class III semaphorins. Neuron 1998; 21:1283-1290.
27. Giger RJ, Cloutier JF, Sahay A et al. Neuropilin-2 is required in vivo for selective axon guidance responses to secreted semaphorins. Neuron 2000; 25:29-41.
28. Chen H, Bagri A, Zupicich JA et al. Neuropilin-2 regulates the development of selective cranial and sensory nerves and hippocampal mossy fiber projections. Neuron 2000; 25:43-56.
29. Takahashi T, Nakamura F, Jin Z et al. Semaphorins A and E act as antagonists of neuropilin-1 and agonists of neuropilin-2 receptors. Nat Neurosci 1998; 1:487-493.
30. Koppel AM, Raper JA. Collapsin-1 covalently dimerizes, and dimerization is necessary for collapsing activity. J Biol Chem 1998; 273:15708-15713.
31. Klostermann A, Lohrum M, Adams RH et al. The chemorepulsive activity of the axonal guidance signal semaphorin D requires dimerization. J Biol Chem 1998; 273:7326-7331.
32. Tamagnone L, Artigiani S, Chen H et al. Plexins are a large family of receptors for trans-membrane, secreted, and GPI-anchored semaphorins in vertebrates. Cell 1999; 99:71-80.
33. Takahashi T, Fournier A, Nakamura F et al. Plexin-neuropilin-1 complexes form functional semaphorin-3A receptors. Cell 1999; 99:59-69.
34. Takahashi T, Strittmatter SM. Plexina1 autoinhibition by the plexin sema domain. Neuron 2001; 29:429-439.
35. Rohm B, Ottemeyer A, Lohrum M et al. Plexin/neuropilin complexes mediate repulsion by the axonal guidance signal semaphorin 3A. Mech Dev 2000; 93:95-104.
36. Cheng HJ, Bagri A, Yaron A et al. Plexin-A3 mediates semaphorin signaling and regulates the development of hippocampal axonal projections. Neuron 2001; 32:249-263.
37. Tessier-Lavigne M. Eph receptor tyrosine kinases, axon repulsion, and the development of topographic maps. Cell 1995; 82:345-348.
38. Keynes R, Tannahill D, Morgenstern DA et al. Surround repulsion of spinal sensory axons in higher vertebrate embryos. Neuron 1997; 18:889-897.

39. Cloutier JF, Giger RJ, Koentges G et al. Neuropilin-2 mediates axonal fasciculation, zonal segregation, but not axonal convergence, of primary accessory olfactory neurons. Neuron 2002; 33:877-892.
40. Bagnard D, Chounlamountri N, Puschel AW et al. Axonal surface molecules act in combination with semaphorin 3a during the establishment of corticothalamic projections. Cereb Cortex 2001; 11:278-285.
41. Fu SY, Sharma K, Luo Y et al. SEMA3A regulates developing sensory projections in the chicken spinal cord. J Neurobiol 2000; 45:227-236.
42. Puschel AW, Adams RH, Betz H. The sensory innervation of the mouse spinal cord may be patterned by differential expression of and differential responsiveness to semaphorins. Mol Cell Neurosci 1996; 7:419-431.
43. Behar O, Golden JA, Mashimo H et al. Semaphorin III is needed for normal patterning and growth of nerves, bones and heart. Nature 1996; 383:525-528.
44. Renzi MJ, Wexler TL, Raper JA. Olfactory sensory axons expressing a dominant-negative semaphorin receptor enter the CNS early and overshoot their target. Neuron 2000; 28:437-447.
45. Zou Y, Stoeckli E, Chen H et al. Squeezing axons out of the gray matter: A role for slit and semaphorin proteins from midline and ventral spinal cord. Cell 2000; 102:363-375.
46. Henke-Fahle S, Beck KW, Puschel AW. Differential responsiveness to the chemorepellent Semaphorin 3A distinguishes ipsi- and contralaterally projecting axons in the chick midbrain. Dev Biol 2001; 237:381-397.
47. Marin O, Yaron A, Bagri A et al. Sorting of striatal and cortical interneurons regulated by semaphorin-neuropilin interactions. Science 2001; 293:872-875.
48. Polleux F, Giger RJ, Ginty DD et al. Patterning of cortical efferent projections by semaphorin-neuropilin interactions. Science 1998; 282:1904-1906.
49. Bagnard D, Thomasset N, Lohrum M et al. Spatial distributions of guidance molecules regulate chemorepulsion and chemoattraction of growth cones. J Neurosci 2000; 20:1030-1035.
50. Sugimoto Y, Taniguchi M, Yagi T et al. Guidance of glial precursor cell migration by secreted cues in the developing optic nerve. Development 2001; 128:3321-3330.
51. de Castro F, Hu L, Drabkin H et al. Chemoattraction and chemorepulsion of olfactory bulb axons by different secreted semaphorins. J Neurosci 1999; 19:4428-4436.
52. Song H, Ming G, He Z et al. Conversion of neuronal growth cone responses from repulsion to attraction by cyclic nucleotides. Science 1998; 281:1515-1518.
53. Polleux F, Morrow T, Ghosh A. Semaphorin 3A is a chemoattractant for cortical apical dendrites. Nature 2000; 404:567-573.
54. Nakamura F, Kalb RG, Strittmatter SM. Molecular basis of semaphorin-mediated axon guidance. J Neurobiol 2000; 44:219-229.
55. Nakamura F, Tanaka M, Takahashi T et al. Neuropilin-1 extracellular domains mediate semaphorin D/III-induced growth cone collapse. Neuron 1998; 21:1093-1100.

THE ROLE OF NEUROPILIN IN VASCULAR AND TUMOR BIOLOGY

Michael Klagsbrun[1,2], Seiji Takashima[3] and Roni Mamluk[1]

SUMMARY

Neuropilin-1 (NRP1) and NRP2 are related transmembrane receptors that function as mediators of neuronal guidance and angiogenesis. NRPs bind members of the class 3 semaphorin family, regulators of neuronal guidance, and of the vascular endothelial growth factor (VEGF) family of angiogenesis factors. There is substantial evidence that NRPs serve as mediators of developmental and tumor angiogenesis. NRPs are expressed in endothelial cells (EC) and bind $VEGF_{165}$. NRP1 is a co-receptor for VEGF receptor-2 (VEGFR2) that enhances the binding of $VEGF_{165}$ to VEGFR2 and $VEGF_{165}$-mediated chemotaxis. NRP1 expression is regulated in EC by tumor necrosis factor-α, the transcription factors dHAND and Ets-1, and vascular injury. During avian blood vessel development NRP1 is expressed only in arteries whereas NRP2 is expressed in veins. Transgenic mouse models demonstrate that NRP1 plays a critical role in embryonic vascular development. Overexpression of NRP1 results in the formation of excess capillaries and hemorrhaging. NRP1 knockouts have defects in yolk sac, embryo and neuronal vascularization, and in development of large vessels in the heart. Tumor cells express NRPs and bind $VEGF_{165}$. NRP1 upregulation is positively correlated with the progression of various tumors. Overexpression of NRP1 in rat tumor cells results in enlarged tumors and substantially enhanced tumor angiogenesis. On the other hand, soluble NRP1 (sNRP1) is an antagonist of tumor angiogenesis. Semaphorin 3A binds to EC and tumor cells. It also inhibits EC motility and capillary sprouting in vitro. $VEGF_{165}$

Departments of [1]Surgical Research and [2]Pathology, Children's Hospital and Harvard Medical School, Boston, MA 02115; and, Department of [3]Internal Medicine and Therapeutics, Osaka University Graduate School of Medicine, Suita Osaka 565-0871, Japan

and Sema3A are competitive inhibitors for NRP1 mediated functions in EC and neurons. These results suggest that NRP1 is a novel regulator of the vascular system.

INTRODUCTION

Neuropilins (NRPs) are mediators of neuronal guidance and angiogenesis.[1-6] NRP1, a 130-140 kDa cell-surface glycoprotein, was first identified in developing nervous tissue.[7-9] It is a highly conserved type 1 membrane protein. Subsequently a second gene, NRP2, was identified that shared a similar structure.[10] Despite a 45-50% structural homology, NRP1 and NRP2 differ considerably in their biological properties (Table 1).

In the nervous system NRP expression is localized to axons as opposed to the somata of neurons. It is expressed in axons of particular neuron classes and at stages when axons are actively growing to form neuronal connections. These original observations suggested that NRP is involved in growth, nerve fiber fasciculation and neuronal guidance (see chapter 2). Subsequently, it was discovered that NRP expression is not confined to the developing embryonic nervous system. It is also expressed in the following: developing heart, vasculature, and limb;[11] many adult tissues such as the heart, placenta, lung, kidney and epidermis,[12-14] and uterine glandular epithelium;[15] and many cell types such as endothelial cells (EC),[11] tumor cells,[12] neural crest cells,[16] osteoblasts,[17,18] marrow stromal cells,[19] human mesangial cells,[20] neuroendocrine cells (NRP2)[21] and glomerular epithelial cells.[22] These expression patterns suggest that NRPs have physiological roles well beyond mediating neuronal guidance. Expression of NRPs in EC and tumor cells will be described in detail below.

NRPs are highly conserved among vertebrate species. The homology between NRP1 and NRP2 is 45%.[10] The primary structure of NRPs contains a relatively large extracellular domain of about 860 amino acids, a transmembrane domain and a relatively short cytoplasmic domain of 40 amino acids.[7,8] The extracellular domain in turn is composed of five subdomains, each of which is thought to be involved in molecular and/or cellular interactions. These subdomains are referred to as a1, a2, b1, b2, and c. The a1a2 and b1b2 are tandem repeats which are involved in ligand binding. The c domain is responsible for homo- or hetero-dimerization of NRP1 and NRP2. The function of the short cytoplasmic domain, the most highly conserved domain, with over 90% homology, is not clear. However, a PDZ domain-containing protein has been isolated using the two-yeast hybrid system that interacts with the C-terminal three amino acids of NRP1 (S-E-A-COOH).[23]

In addition to full-length NRPs, some cell types also express truncated NRP isoforms.[13,14] These proteins contain the a1a2 and b1b2 subdomains, but lack the c, transmembrane and cytoplasmic domains. These naturally occurring 60-90 kDa proteins are soluble and released by cells. Several soluble NRPs (sNRP) have been cloned, three sNRP1s and one sNRP2. The sNRP molecules are produced by premature truncation within introns and as a result are characterized by having intron-derived sequences, nucleotides and amino acids, at their C-termini.

Table 1. Comparison of NRP1 and NRP2

	NRP1	NRP2	References
Chromosome	10p12	2q34	24
Isoforms	NRP1	NRP2a, NRP2b	14,26
Vascular Expression	Arterial EC	Venous EC	33,34
VEGF Family Ligands			
$VEGF_{165}$	+	+	12
$VEGF_{121}$	-	-	12
$VEGF_{145}$	-	+	39
VEGF-B	+	?	41
VEGF-C	?	+	44
VEGF-E	+	-	42
PlGF-2	+	-	40
Activation by Semaphorins			
Sema3A	+	-	4
Sema3F	-	+	4
Knockout:			
Lethality	E 12.5 –13.5	Viable	61-64
Vasculature	Severely impaired in: yolk sac, embryo, nervous system, Heart	Normal	

The two NRP genes, NRP1 and NRP2 map to chromosomes 10p12 and 2q34, respectively.[14,24] These two genes span over 120 and 112 kb, respectively, and are composed of 17 exons. Five of the exons are identical in size, suggesting that they arose by gene duplication. The NRP2 gene expresses several alternatively spliced variants, for example divergent NRP2 cytoplasmic domains. These splice variants are expressed in a variety of tissues, mostly in a non-overlapping manner (See chapter 5 for more details).

NRPs are receptors for members of the class 3 semaphorin family, regulators of neuronal guidance[10,25] and for the VEGF family of angiogenesis factors.[12] There are six Class 3 semaphorins, which bind to NRP1 and NRP2 with different specificities.[26-29] Semaphorin 3A (Sema3A), the best characterized semaphorin, repels axons, collapses growth cones of dorsal root ganglion neurons and regulates migration of cortical neurons in an NRP1-dependent manner.[30,31] NRPs do not appear to directly activate signaling pathways in neurons. Instead, signaling is mediated by the interactions of a Sema3A/NRP1 complex with plexins which are transmembrane signaling receptors.[28,32-34] NRP1/plexin complex formation enhances Sema3A binding to NRP1. L1-CAM, a neuronal adhesion molecule has also been demonstrated to be a component of the Sema3A receptor complex.[35] (See chapters 6 and 8 for more details).

VEGF is the predominant regulator of developmental and tumor angiogenesis.[2,36,37] VEGF activities are mediated via three receptor tyrosine kinases (RTK), VEGFR1, VEGFR2 and VEGFR3. NRPs are novel receptors for $VEGF_{165}$ but do not appear to be receptor tyrosine kinases.[12,38] Therefore, they constitute a second class of VEGF receptors. Binding of VEGF to NRP is isoform specific (Table 1). $VEGF_{165}$, but not $VEGF_{121}$, binds NRP1 because $VEGF_{121}$ lacks the domain encoded by exon 7 that is responsible for NRP binding.[12,38] Exon 7 is also present in $VEGF_{189}$ suggesting its binds to NRP1. There is a degree of specificity in NRP binding (Table 1). For example $VEGF_{145}$ binds NRP2 but not NRP1[39] and placenta growth factor-2 (PlGF2) binds NRP1 but not NRP2.[40] Other members of the VEGF family, VEGF-B[41] and VEGF-E[42,43] are also ligands for NRP1. A recent report has demonstrated that VEGF-C binds to NRP2.[44] Although NRPs do not appear to be tyrosine kinases they may contribute to signaling by interactions with VEGFR1 and/or VEGFR2.[45-47] (see chapter 7 for more details)

NEUROPILIN EXPRESSION IN ENDOTHELIAL CELLS

There have been a number of reports demonstrating that blood vessel EC express NRPs; for example, umbilical vein EC, aortic EC and capillary EC.[12,48] Expression *in vivo* has been demonstrated, for example, in the heart,[11] coronary blood vessels,[49] glomerular capillaries,[50] and bone capillaries.[17,18] Vascular smooth muscle cells (vSMC) also express NRP1, which suggests a possible role in EC/vSMC interactions.[51]

NRPs are not ubiquitously expressed in EC. Previous reports have shown differential growth factor/receptor expression patterns on blood vessels; for example, expression of ephrinB2 on arteries and of EphB receptors on veins.[52] Two recent reports demonstrated that there are differential embryonic blood vessel expression patterns for NRP1 and NRP2 as well.[53,54] In the avian vascular system NRP1 and NRP2 are both expressed in blood islands which are the earliest vascular structures. However, once arteries and veins differentiate, NRP1 is expressed exclusively in arteries and in mesenchyme surrounding developing arteries,[53,54] and NRP2 is expressed only in veins.[54] Similar expression patterns were detected throughout chick and quail development. Using quail arterial EC grafted onto chick embryos, EC expressing specific markers colonized both the arteries and veins of the chick embryo,[53] suggesting that expression of NRPs in EC is insufficient to determine the fate of these cells. In mice, a recent report has shown that in the retina, NRP1 is predominantly expressed in arterioles, suggesting there may be similar NRP expression pattern in mammals as in birds.[55]

REGULATION OF NEUROPILIN EXPRESSION IN BLOOD VESSELS

NRP expression is regulated by cytokines and transcription factors. For example, tumor necrosis factor-α (TNF-α1 up-regulates in a dose- and time-dependent manner the expression and the function of VEGFR2, as well as the expression of NRP1 in human EC.[56]

The basic helix-loop-helix transcription factor, dHAND/Hand2, is expressed in the developing vascular mesenchyme and derivative vSMC. Targeted deletion of the dHAND gene in mice revealed severe defects of embryonic and yolk sac vascular development by E9.5.[57] In the dHAND mouse knockout, EC appear to be normal. The vascular mesenchymal cells migrated appropriately, but failed to make contact with vascular EC and did not differentiate into vSMC. In a subtractive hybridization screen for genes comparing wild type and dHAND-null hearts, NRP1 was found to be downregulated in dHAND mutants. At E9.5 the expression of dHAND and NRP1 in wild type mice overlapped in the developing vasculature, for example, in aorta, aortic arch arteries and yolk sac. In the dHAND-null mice NRP1 expression was severely downregulated, specifically in blood vessels that expressed dHAND. These results suggest that dHAND is required for normal cardiovascular development and that it regulates angiogenesis, possibly through a NRP1 dependent mechanism.

Whereas dHAND regulates NRP1 expression in vSMC, another transcription factor, Ets-1, induces NRP1 expression in EC. Ets-1 is expressed in EC during angiogenesis and is induced by angiogenesis factors including VEGF.[58] Ets-1 was transiently overexpressed in human umbilical vein EC (HUVEC) and potential downstream targets of Ets-1 were analyzed by cDNA microarray analysis.[59] NRP1 was one of several angiogenesis-related genes induced by Ets-1. In contrast, dominant negative Ets-1 decreased the levels of NRP1 mRNA and protein. Since TNF-α increases the expression of both NRP1[56] and Ets-1[60] in EC, it is possible that TNF-α induces NRP1 expression via Ets-1.

NEUROPILIN AND ANGIOGENESIS

NRP1 appears to be a co-receptor of VEGFR2 in cultured EC.[12] When coexpressed in cells with VEGFR2, NRP1 enhances the binding of $VEGF_{165}$ to VEGFR2 and $VEGF_{165}$-mediated chemotaxis. Conversely, inhibition of $VEGF_{165}$ binding to NRP1 inhibits its binding to VEGFR2 and its mitogenic activity for EC.

There is ample evidence from transgenic mouse studies that NRPs mediate angiogenesis, both normal and pathological. The first hint to this effect was a transgenic mouse study in which NRP1 was overexpressed.[11] In wild type mice, NRP1 is expressed in the cardiovascular system, nervous system and limbs at particular developmental stages. The transgenics overexpressing NRP1 were embryonic lethal and displayed several morphological abnormalities. Besides ectopic sprouting and defasciculation of nerve fibers, there was an abnormal vascular phenotype that included excess capillaries and blood vessels, dilation of blood vessels, hemorrhaging and malformed hearts. The chimeric embryos usually appeared redder than their normal counterparts, suggesting that blood vessels were leaky which was possibly due to the enhanced vascular permeability activity of $VEGF_{165}$. Extra digit formation was also noted. These abnormalities occurred in the organs in which NRP1 was expressed in normal development. It was concluded that expression of NRP1 was essential not only for neuronal development but also development of the cardiovascular system and limbs.

Knockout studies have been very useful in determining the physiological role of NRPs in angiogenesis. In the initial study it was demonstrated that NRP1-deficient mutant mice were embryonic lethal between E12.5 to E13.5 and had, for example, severe abnormalities in the trajectory of efferent fibers of the peripheral nervous system.[61] Interestingly, it was mentioned but not demonstrated that the embryo died due to cardiovascular defects. A follow up study analyzed cardiovascular defects in depth.[62] The NRP1 mutant mouse embryos exhibited defects in yolk sac, embryo and neuronal vascularization, and in development of large vessels in the heart. In yolk sacs and embryos the vascular network of large and small vessels was disorganized, the capillary networks were sparse, and normal branching did not occur. In the central nervous system (CNS) capillary invasion into the CNS was delayed for more than 1 day and the capillary networks that were in the CNS were disorganized and had degenerated. In the cardiovascular system the mutant embryos showed abnormal development, such as agenesis of the branchial arch-related great vessels and dorsal aorta and transposition of the aortic arch. For example, the most frequent variant was the absence of the left 4th branchial arch artery. The development of heart outflow tracts was also disturbed and separation of the truncus arteriosus was incomplete (persistent truncus arteriosus).

On the other hand, two reports on NRP2 knockouts did not report any abnormal vascular phenotype.[63,64] Unlike the NRP1 knockouts, NRP2 mutant mice were viable into adulthood. NRP2 was required for the organization and fasciculation of cranial nerves and spinal nerves and for Sema3F activity, but possible effects on the cardiovascular system were not described.

Double knockouts in which both NRP1 and NRP2 were targeted ($NRP1^{-/-}$/$NRP2^{-/-}$) have also been generated.[65] These mice died in utero at E8. Their yolk sacs showed an absence of branching arteries and veins, the absence of a capillary bed and the presence of large avascular spaces between the blood vessels. The embryos had large avascular regions in the head and trunk, and blood vessel sprouts that were not connected. These double NRP1/NRP2 knockout mice had an even more severely abnormal vascular phenotype than either NRP1 or NRP2 single knockouts. Their abnormal vascular phenotype resembled those of VEGF and VEGFR2 knockouts. These results suggest that NRPs are early genes in embryonic vessel development and that both NRP1 and NRP2 are involved in normal blood vessel development.

NRP1 knockout embryos have been used to analyze NRP1-dependent vascular function in vitro as well as in vivo. Cultured wild type para-aortic splanchnopleural mesoderm (P-Sp) explants supported vasculogenesis and angiogenesis whereas P-Sp explants derived from $NRP1^{-/-}$ mice had defects in capillary sprouting in vitro, consistent with the impaired vascular sprouting demonstrated in vivo in the CNS and cardiovascular system.[66] A soluble NRP1 (sNRP1), corresponding to the a1a2, b1b2 and c extracellular domains of NRP1, inhibited capillary sprouting in the cultured wild-type P-Sp explants. In contrast, an sNRP1 dimer produced by fusion with the Fc part of human IgG, enhanced vascular development in wild type explants and rescued the defective vascular phenotype of mutant $NRP1^{-/-}$ explants. Furthermore, sNRP-Fc dimer, when injected into pregnant mice, reversed and rescued the $NRP1^{-/-}$ embryo phenotype. sNRP1 monomers have been shown to bind $VEGF_{165}$ and inhibit VEGF mitogenic activity for EC.[13] Whereas an sNRP1 monomer appears to sequester $VEGF_{165}$ and inhibit its activity, sNRP1 dimer appears to deliver $VEGF_{165}$ to EC VEGFR2, thereby promoting angiogenesis and vasculogenesis.

Semaphorins were first described as mediators of neuronal guidance acting via NRPs[10,25] but they may also be mediators of EC activity. Sema3A binds to aortic EC and inhibits the motility of EC only if they express NRP1.[48] Sema3A also inhibits the capillary sprouting of EC from rat aortic ring segments in an in vitro angiogenesis assay. $VEGF_{165}$ and Sema3A are competitive inhibitors in EC motility, ligand binding and dorsal root ganglia collapse assays, suggesting possible overlapping binding sites.[48] $VEGF_{165}$ and Sema3A are also antagonists in neuronal survival/ apoptosis assays.[67] $VEGF_{165}$ interacts with neuronal NRP1 and is a survival factor for neurons, such as hippocampal neurons and motor neurons,[68-70] whereas Sema 3A induces neuronal apoptosis.[67,71] These results suggest that a balance of semaphorins and $VEGF_{165}$ can modulate the migration, apoptosis/survival and proliferation of neurons and EC through shared receptors.

TUMOR CELL NEUROPILIN

Many tumor cell types express NRP1 and NRP2 and bind $VEGF_{165}$. The first report was that of $VEGF_{165}$ binding to PC3 prostate and MDA-MB-231 breast carcinoma cells.[12] $VEGF_{165}$ binds to NRP1 in these tumor cell types with a Kd of approximately 2×10^{-10} M, with about $1\text{-}2 \times 10^5$ receptors per cell. NRPs are the only VEGF receptors expressed by these tumor cells so that any $VEGF_{165}$ activity for tumor cells is mediated by NRPs. Subsequently a number of tumor cell types have been shown to express NRP1 and/or NRP2 in vitro and in vivo. These include prostate carcinoma,[72] melanoma,[73] astrocytoma,[74] osteosarcoma[75] and rat pituitary tumors.[76] In several clinical studies NRP1 and NRP2 expression was correlated with increased aggressiveness, malignancy or hypervascularity. For example, NRP1 was upregulated in primary sporadic prostate tumors at different clinical stages as determined by quantitative RT-PCR.[72] The correlation between NRP1 overexpression with advanced disease and a high Gleason grade (a morphological measure of prostate cancer progression) suggested that NRP1 overexpression might be a marker of aggressiveness. The expression pattern of NRP1 and VEGF by human astrocytoma cell lines and specimens was closely correlated and associated with malignant astrocytomas.[74] Osteosarcoma, a malignant bone tumor characterized by hypervascularity, expressed NRP2 (24 out of 30 specimens) and the NRP2-positive tumors showed both a significantly increased vascularity and a significantly poorer prognosis than those without NRP2.[75]

Besides these correlational studies, the function of NRP1 in tumor cells had been analyzed more directly. NRP1 was overexpressed in Dunning rat prostate carcinoma AT2.1 cells using a tetracycline-inducible promoter.[77] Concomitant with increased NRP1 expression in response to a tetracycline homologue, doxycycline (Dox), AT2.1 cell migration was enhanced, and $VEGF_{165}$ binding was increased 3- to 4- fold in vitro. However, induction of NRP1 did not affect tumor cell proliferation. When rats injected with AT2.1/NRP1 tumor cells were fed Dox, NRP1 synthesis was induced in vivo, and tumor size was increased 2.5- to 7-fold, in a three to four week period, compared to control. The larger tumors with induced NRP1 expression were characterized by markedly increased microvessel density, increased proliferating EC, dilated blood vessels and notably less tumor cell apoptosis compared to non-induced controls. It was concluded that NRP1 expression results in enlarged tumors associated with substantially enhanced tumor angiogenesis.

On the other hand, sNRP1 is a tumor antagonist.[13] Tumors of rat prostate carcinoma cells overexpressing recombinant sNRP1 in vivo were characterized by extensive hemorrhage, damaged vessels and apoptotic tumor cells. Since sNRP1 inhibits ^{125}I-$VEGF_{165}$ binding to EC and $VEGF_{165}$-induced tyrosine phosphorylation of VEGFR2 in EC in vitro, this tumor phenotype may be due to $VEGF_{165}$ withdrawal and lack of bioavailability. Withdrawal of $VEGF_{165}$ from tumors using a Tet-off system has previously been shown to result in vascular damage, EC apoptosis, hemorrhage and extensive tumor necrosis.[78]

NRPs may also be involved in tumor cell survival.[79] Suppression of VEGF expression in metastatic breast carcinoma MDA-MB-231 cells in vitro induced apoptosis. These effects were probably NRP-dependent since NRP1 and NRP2 are the only VEGF receptors expressed in these cells. Furthermore, $VEGF_{165}$ enhanced breast carcinoma cell survival but $VEGF_{121}$, the isoform which lacks the ability to bind to NRP1 did not, implicating a role for NRP1 in tumor cell survival.

Expression of several class 3 semaphorins has been studied extensively in lung cancer. The progression of small cell lung cancer correlates with a deletion in 3p21, and a loss of semaphorin expression; in particular, Sema3B and Sema3F.[80-82] In non small cell lung carcinomas, low levels of Sema3F expression correlated with higher stages of disease.[83] Sema3B and Sema3F were transfected into lung cancer NCI-H1299 cells, which do not express either gene.[84] Colony formation of H1299 cells was reduced by 90% after transfection with wild type Sema3B as compared with the control vector. A 30-40% reduction in colony formation was seen after the transfection of Sema3F or Sema3B variants carrying single amino acid missense mutations that had been associated with lung cancer. H1299 cells transfected with wild type, but not mutant Sema3B, underwent apoptosis. Lung cancers ($n = 34$) always expressed NRP1, and most of these expressed NRP2. In a very recent report, ovarian tumor cells overexpressing Sema3B exhibited diminished tumorigenicity in mice. Taken together, these results suggest that Sema3B and Sema3A are functional tumor suppressor genes.[85]

On the other hand, Sema3C mRNA is overexpressed several-fold in metastatic lung tumors as determined by differential display and Northern blot analysis of lung tumor cell lines.[86] Thus, class 3 semaphorins are involved in tumor progression and metastasis, both as inhibitors and promoters.

VASCULAR INJURY

NRPs are induced following injury in several model systems; for example, cerebral artery occlusion, cerebral ischemia, hind limb ischemia and retinal vascularization. A recurring pattern is that NRPs are highly expressed in the developing embryo as compared with the normal adult, but are induced following injury or ischemia. Several diseases characterized by increased angiogenesis, such as diabetic retinopathy and rheumatoid arthritis, show NRP1 upregulation. The first demonstration that NRP1 is induced following injury was in regenerating Xenopus optic nerves.[87] In embryos NRP1 was expressed in retinal ganglion cells, maximal at stages 41-43, and then decreased as the tadpole developed. After stage 50 NRP1 expression was almost nil. When the tadpole optic nerves were crushed and prompted to regenerate, however, NRP protein reappeared in the optic nerve fibers, being maximal at the second and third week after the optic nerve crush, and then declined thereafter.

Ischemia upregulates NRP expression. In the adult mouse, ischemic brain NRP1 mRNA expression was significantly up-regulated as early as two hours and persisted at least 28 days after focal cerebral ischemia.[88] Acute up-regulation of NRP1 mRNA was primarily localized to the ischemic neurons but there was also a marked increase in NRP1 expression in EC of cerebral blood vessels at the border and in the core of the ischemic lesion seven days after ischemia. NRP1 expression persisted on these vessels for at least 28 days after ischemia. Activated astrocytes also exhibited NRP1 immunoreactivity during 7 to 28 days of ischemia. Double immunofluorescent staining showed colocalization of NRP1 and VEGF to cerebral blood vessels and activated astrocytes. These results suggest that in addition to its role in axonal growth, up-regulation of NRP1 may contribute to neovascular formation in the adult ischemic brain.

In another mouse ischemia model system very little NRP2 expression was observed in normal blood vessels after birth (Takashima et. al, unpublished). However, it was possible to induce NRP2 expression in blood vessels in response to ischemia in a hind limb model in which occlusion of the femoral artery by ligation resulted in the sprouting of new vessels (Fig. 1). Prior to injury or in a sham operation without ligation, NRP2 was not expressed in the femoral artery. However, after one week, NRP2 expression was clearly seen in the sprouting vessels at the edge of the ligated artery. By two weeks NRP2 expression was more prominent and was detected in EC in the vascular wall of newly developed mid-sized arteries. These expression profiles indicate that NRP2 is expressed primarily in the embryo and

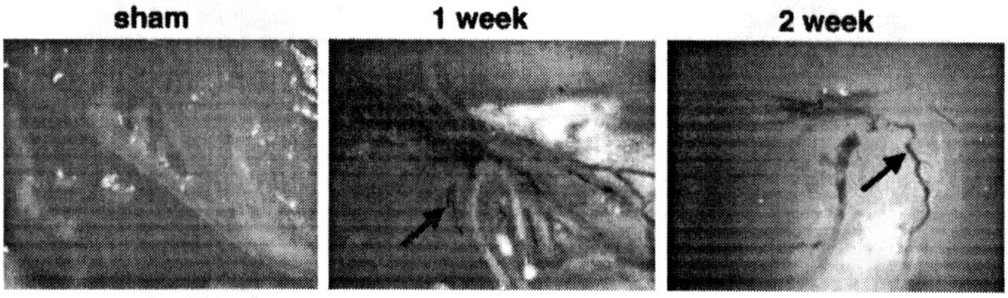

Figure 1. NRP2-LacZ expression in an 8-10 week adult ischemic hindlimb.[91] Ischemia was induced by ligation of the femoral artery as previously described. Left: Two weeks after a sham operation without ligation of the femoral artery, there was very little if any NRP2 expression. Center: At one week after inducing ischemia, small sprouting capillaries surrounding the femoral artery expressed NRP2 at the site of the artery where vessel ligation occurred, (arrow). Right: At two weeks after inducing ischemia, NRP2 was expressed in mid-sized vessels growing from the cutting edge of the injured vessels and in small sprouting vessels (arrow). A color version of this figure is included in the insert that precedes Chapter 1.

extra-embryonic tissue, whereas expression in the adult is atypical and occurs only when induced, for example, by ischemia.

NRP1 expression is also induced in retinal neovascularization. A model of retinopathy of prematurity (ROP) was produced by ischemia induced ocular neovascularization. Postnatal day-7 mice were exposed to 75% oxygen for five days and then returned to room air for five days.[89] Retinal neovascularization was visualized by injection of fluorescein-dextran. Expression of NRP1 and VEGFR2 mRNAs was colocalized in the area of neovascularization. In addition, expression of VEGFR2 and NRP1 was restricted to neovascularized vessels of the retina from ROP mice. The restricted expression of VEGFR2 and NRP1 on neovascularized vessels suggests that these molecules may play important roles in retinal neovascularization.

In a clinical study, fibrovascular tissues were obtained at vitrectomy from 22 cases with proliferative diabetic retinopathy.[90] RT-PCR analysis demonstrated the expression of VEGF receptors VEGFR1, VEGFR2 and NRP1 in 12, 14 and 14 of 22 cases, respectively. Notably, VEGFR2 and NRP1 were simultaneously expressed in the identical 14 tissues. The vascular density of fibrovascular tissues as determined by immunohistochemistry for CD34, an EC marker, was significantly higher in cases with the expression of VEGFR2 and NRP1 versus those without their expression. VEGFR1 expression had no such relationship with the vascular density. It was concluded that coexpression of VEGFR2 and NRP1 may facilitate fibrovascular proliferation in diabetic retinopathy.

PERSPECTIVES AND FUTURE DIRECTIONS

There is substantial evidence, based on cell culture and transgenic mouse studies which indicates that NRPs are novel and significant regulators of blood vessel development. In embryonic development NRP expression occurs early, in the blood islands of the yolk sac. NRP expression is required for the normal branching and organization of large vessels and capillaries in the developing yolk sac and embryo. In the developing heart NRP1 is required for normal formation of the large arteries. In the adult, NRP expression is generally reduced but it is strongly upregulated in blood vessels in response to vascular injury. How NRP expression in EC affects cellular function is not clear. One can speculate that the ability of NRPs to bind VEGF and class 3 semaphorins to EC must play a role and that in the absence of NRPs, VEGF and semaphorin activity is compromised. Since both VEGF and Sema3A are involved in forming networks of blood vessels and neurons, respectively, these processes might be adversely affected by a lack of NRP expression.

There are two NRPs, NRP1 and NRP2, and they probably are responsible for some non-overlapping functions (Table 1). For example, the abnormal vascular phenotype is much more severe in NRP1-deficient mice than in NRP2-deficient mice. During development, NRP1 is expressed by arteries and NRP2 by veins. There are some differences in NRP1 and NRP2 ligand interactions. For example, NRP1 is activated by Sema3A and NRP2 by Sema3F. PlGF2 binds NRP1 but not NRP2, whereas $VEGF_{145}$ binds NRP2 but not NRP1. Thus, arteries and veins may bind different ligands and thereby may be subjected to different signals. Differential interactions of NRP1 and NRP2 with VEGFR1 and/or VEGFR2 might also contribute to different signaling pathways in arteries and veins.

Tumor cells are among the highest expressers of NRPs and as a consequence bind VEGF, typically in the absence of other VEGF receptors. The significance of direct VEGF binding to tumor cells is unknown but might involve enhancement of tumor cell migration and survival. NRP1 overexpression in tumor cells enhances tumor angiogenesis whereas sNRP1 suppresses it. The reason for this might be VEGF bioavailability. Full length NRP is membrane anchored and might be expected to concentrate VEGF on the cell surface, thereby signaling neighboring EC or the tumor cells themselves. On the other hand, sNRPs are soluble and may sequester VEGF away from the cell surface, thereby inducing cell apoptosis. Tumor cells express both full-length NRP and sNRP, thus, a balance of these two types of NRP might contribute to the level of tumor angiogenesis

Future Directions include: 1) determination of the mechanisms by which NRPs regulate angiogenesis, for example, whether NRP is involved in ligand signaling, directly or as co-receptors for VEGF RTKs, Plexins or other receptors; 2) identification of upstream regulators of NRP1 function. So far TNF-α, and the transcription factors dHAND and Ets-1 have been implicated.; 3) identification of target genes and proteins downstream of NRP; 4) determination of novel VEGF and semaphorin biological functions given that many different cell types besides neuronal cells and EC express NRP and bind these two families of ligands; and 5) determination of whether there are any NRP clinical applications. Possible tumor antagonists include sNRP which induces tumor cell apoptosis, Sema3A, which blocks in vitro angiogenesis, and Sema3B and Sema3F which may have tumor suppressor activity.

ACKNOWLEDGMENTS

This article was supported by NIH grants CA37392 and CA44548 (MK) and a grant from the Erenst Schering Research Foundation in Berlin, Germany (RM). We thank Alexandra Grady for preparation of the manuscript.

REFERENCES

1. Fujisawa H, Kitsukawa, T. Receptors for collapsin/semaphorins. Curr Opin Neurobiol 1998; 8:587-592.
2. Neufeld G, Cohen T, Gengrinovitch S et al. Vascular endothelial growth factor (VEGF) and its receptors. FASEB J 1999; 13:9-22.
3. Shima D, Mailhos C. Vascular development biology: getting nervous. Curr Opin Genet Dev 2000; 10:536-542.
4. Raper JA. Semaphorins and their receptors in vertebrates and invertebrates. Curr Opin Neurobiol 2000; 10:88-94.
5. Miao HQ, Klagsbrun M. Neuropilin is a mediator of angiogenesis. Cancer Metastasis Rev 2000; 19:29-37.
6. Soker S. Neuropilin in the midst of cell migration and retraction. Int J Oncol 2001; 33:433-437.
7. Takagi S, Kasuya Y, Shimizu M et al. Expression of a cell adhesion molecule, neuropilin, in the developing chick nervous system. Dev Biol 1995; 170:207-222.
8. Kawakami A, Kitsukawa T, Takagi S et al. Developmentally regulated expression of a cell surface protein, neuropilin, in the mouse nervous system. J Neurobiol 1996; 29:1-17.
9. Fujisawa H, Kitsukawa T, Kawakami A et al. Roles of a neuronal cell-surface molecule, neuropilin, in nerve fiber fasciculation and guidance. Cell Tissue Res 1997; 290:465-470.
10. Kolodkin AL, Levengood DV, Rowe EG et al. Neuropilin is a semaphorin III receptor. Cell 1997; 90:753-762.
11. Kitsukawa T, Shimono A, Kawakami A et al. Overexpression of a membrane protein, neuropilin, in chimeric mice causes anomalies in the cardiovascular system, nervous system and limbs. Development 1995; 121:4309-4318.
12. Soker S, Takashima S, Miao HQ et al. Neuropilin-1 is expressed by endothelial and tumor cells as an isoform-specific receptor for vascular endothelial growth factor. Cell 1998; 92:735-745.
13. Gagnon ML, Bielenberg DR, Gechtman Z et al. Identification of a natural soluble neuropilin-1 that binds vascular endothelial growth factor: In vivo expression and antitumor activity. Proc Natl Acad Sci USA 2000; 97:2573-2578.
14. Rossignol M, Gagnon ML, Klagsbrun M. Genomic organization of human neuropilin-1 and neuropilin-2 genes: identification and distribution of splice variants and soluble isoforms. Genomics 2000; 70:211-222.
15. Pavelock K, Braas K, Ouafik L et al. Differential expression and regulation of the vascular endothelial growth factor receptors neuropilin-1 and neuropilin-2 in rat uterus. Endocrinology 2001; 142:613-622.
16. Eickholt BJ, Mackenzie SL, Graham A et al. Evidence for collapsin-1 functioning in the control of neural crest migration in both trunk and hindbrain regions. Development 1999; 126:2181-2189.

17. Deckers MM, Karperien M, van der Bent C et al. Expression of vascular endothelial growth factors and their receptors during osteoblast differentiation. Endocrinology 2000; 141:1667-1674.
18. Harper J, Gerstenfeld LC, Klagsbrun M. Neuropilin-1 expression in osteogenic cells: down-regulation during differentiation of osteoblasts into osteocytes. J Cell Biochem 2001; 81:82-92.
19. Tordjman R, Ortega N, Coulombel L et al. Neuropilin-1 is expressed on bone marrow stromal cells: a novel interaction with hematopoietic cells? Blood 1999; 94:2301-2309.
20. Thomas S, Vanuystel J, Gruden G et al. Vascular endothelial growth factor receptors in human mesangium in vitro and in glomerular disease. J Am Soc Nephrol 2000; 11:1236-1243.
21. Cohen T, Gluzman-Poltorak Z, Brodzky A et al. Neuroendocrine cells along the digestive tract express neuropilin-2. Biochem Biophys Res Commun 2001; 284:395-403.
22. Harper SJ, Xing CY, Whittle C et al. Expression of neuropilin-1 by human glomerular epithelial cells in vitro and in vivo. Clin Sci 2001; 101:439-446.
23. Cai H, Reed RR. Cloning and characterization of neuropilin-1-interacting protein: a PSD-95/Dlg/ZO-1 domain-containing protein that interacts with the cytoplasmic domain of neuropilin-1. J Neurosci 1999; 19:6519-6527.
24. Rossignol M, Beggs AH, Pierce EA et al. Human neuropilin-1 and neuropilin-2 map to 10p12 and 2q34, respectively. Genomics 1999; 57:459-460.
25. He Z, Tessier-Lavigne M. Neuropilin is a receptor for the axonal chemorepellent Semaphorin III. Cell 1997; 90:739-751.
26. Chen H, Chedotal A, He Z et al. Neuropilin-2, a novel member of the neuropilin family, is a high affinity receptor for the semaphorins Sema E and Sema IV but not Sema III. Neuron 1997; 19:547-559.
27. Chen H, He Z, Bagri A et al. Semaphorin-neuropilin interactions underlying sympathetic axon responses to class III semaphorins. Neuron 1998; 21:1283-1290.
28. Takahashi T, Nakamura F, Jin Z et al. Semaphorins A and E act as antagonists of neuropilin-1 and agonists of neuropilin-2 receptors. Nat Neurosci 1998; 1:487-493.
29. Giger RJ, Urquhart ER, Gillespie SK et al. Neuropilin-2 is a receptor for semaphorin IV: insight into the structural basis of receptor function and specificity. Neuron 1998; 21:1079-1092.
30. Luo Y, Raible D, Raper JA. Collapsin: a protein in brain that induces the collapse and paralysis of neuronal growth cones. Cell 1993; 75:217-227.
31. Marin O, Yaron A, Bargri A et al. Sorting of striatal and cortical interneurons regulated by semaphorin-neuropilin interactions. Science 2001; 293:872-875.
32. Winberg ML, Noordermeer JN, Tamagnone L et al. Plexin A is a neuronal semaphorin receptor that controls axon guidance. Cell 1998; 95:903-916.
33. Tamagnone L, Artigiani S, Chen H et al. Plexins are a large family of receptors for transmembrane, secreted, and GPI-anchored semaphorins in vertebrates. Cell 1999; 99:71-80.
34. Tamagnone L, Comoglio PM. Signaling by semaphorin receptors: cell guidance and beyond. Trends Cell Biol 2000; 10:377-383.
35. Castellani V, Chedotal A, Schachne, M et al. Analysis of the L1-deficient mouse phenotype reveals cross-talk between Sema3A and L1 signaling pathways in axonal guidance. Neuron 2000; 27:237-249.
36. Ferrara N. Molecular and biological properties of vascular endothelial growth factor. J Mol Med 1999; 77:527-543.
37. Dvorak HF, Nagy JA, Feng D et al. Vascular permeability factor/vascular endothelial growth factor and the significance of microvascular hyperpermeability in angiogenesis. Curr Top Microbiol Immunol 1999; 237:97-132.
38. Soker S, Fidder H, Neufeld G et al. Characterization of novel vascular endothelial growth factor (VEGF) receptors on tumor cells that bind VEGF165 via its exon 7-encoded domain. J Biol Chem 1996; 271:5761-5767.

39. Gluzman-Poltorak Z, Cohen T, Herzog Y et al. Neuropilin-2 is a receptor for the vascular endothelial growth factor (VEGF) forms VEGF-145 and VEGF-165. J Biol Chem 2000; 275:29922.
40. Migdal M, Huppertz B, Tessler S et al. Neuropilin-1 is a placenta growth factor-2 receptor. J Biol Chem 1998; 273:22272-22278.
41. Makinen T, Olofsson B, Karpanen T et al. Differential binding of vascular endothelial growth factor B splice and proteolytic isoforms to neuropilin-1. J Biol Chem 1999; 274:21217-21222.
42. Wise LM, Veikkola T, Mercer AA et al. Vascular endothelial growth factor (VEGF)-like protein from orf virus NZ2 binds to VEGFR2 and neuropilin-1. Proc Natl Acad Sci USA 1999; 96:3071-3076.
43. Savory LJ, Stacker SA, Fleming SB et al. Viral vascular endothelial growth factor plays a critical role in orf virus infection. J Virol 2000; 74:10699-10706.
44. Karkkainen MJ, Saaristo A, Jussila L et al. A model for gene therapy of human hereditary lymphedema. Proc Natl Acad Sci USA 2001; 98:12677-12682.
45. Fuh G, Garcia KC, de Vos AM. The interaction of neuropilin-1 with vascular endothelial growth factor and its receptor flt-1. J Biol Chem 2000; 275:26690-26695.
46. Whitaker GB, Limberg BJ, Rosenbaum JS. Vascular endothelial growth factor receptor-2 and neuropilin-1 form a receptor complex that is responsible for the differential signaling potency of VEGF(165) and VEGF(121). J Biol Chem 2001; 276:25520-25531.
47. Gluzman-Poltorak Z, Cohen T, Shibuya M et al. Vascular endothelial growth factor receptor-1 and neuropilin-2 form complexes. J Biol Chem 2001; 276:18688-18694.
48. Miao HQ, Soker S, Feiner L et al. Neuropilin-1 mediates collapsin-1/semaphorin III inhibition of endothelial cell motility: functional competition of collapsin-1 and vascular endothelial growth factor-165. J Cell Biol 1999; 146:233-242.
49. Partanen TA, Makinen T, Arola J et al. Endothelial growth factor receptors in human fetal heart. Circulation 1999; 100:583-586.
50. Robert B, Zhao X, Abrahamson DR. Coexpression of neuropilin-1, Flk1, and VEGF(164) in developing and mature mouse kidney glomeruli. Am J Physiol 2000; 279:F275-F282.
51. Ishida A, Murray J, Saito Y et al. Expression of vascular endothelial growth factor receptors in smooth muscle cells. J Cell Physiol 2001; 188:359-368.
52. Adams RH, Wilkinson GA, Weiss C et al. Roles of ephrinB ligands and EphB recceptors in cardiovascular development: demarcation of arterial/venous domains, vascular morphogenesis, and sprouting angiogenesis. Genes Dev 1999; 13:295-306.
53. Moyon D, Pardanaud L, Yuan L et al. Plasticity of endothelial cells during arterial-venous differentiation in the avian embryo. Development 2001; 128:3359-3370.
54. Herzog Y, Kalcheim C, Kahane N et al. Differential expression of neuropilin-1 and neuropilin-2 in arteries and veins. Mech Dev 2001; 109:115-119.
55. Stalmans I, Ng YS, Rohan R et al. Arteriolar and venular patterning in retinas of mice selectively expressing VEGF isoforms. J Clin Invest 2002; 109:327-336.
56. Giraudo E, Primo L, Audero E et al. Tumor necrosis factor-alpha regulates expression of vascular endothelial growth factor receptor-2 and of its co-receptor neuropilin-1 in human vascular endothelial cells. J Biol Chem 1998; 273:22128-22135.
57. Yamagishi H, Olson EN, Srivastava D. The basic helix-loop-helix transcription factor, dHAND, is required for vascular development. J Clin Invest 2000; 105:261-270.
58. Sato Y, Abe M, Tanaka K et al. Signal transduction and transcriptional regulation of angiogenesis. Adv Exp Med Biol 2000; 476:109-115.
59. Teruyama K, Abe M, Nakano T et al. Neurophilin-1 is a downstream target of transcription factor Ets-1 in human umbilical vein endothelial cells. FEBS Lett 2001; 504:1-4.
60. Wernert N, Raes MB, Lassalle P et al. c-ets1 proto-oncogene is a transcription factor expressed in endothelial cells during tumor vascularization and other forms of angiogenesis in humans. Am J Pathol 1992; 140:119-127.

61. Kitsukawa T, Shimizu M, Sanbo M et al. Neuropilin-semaphorin III/D-mediated chemorepulsive signals play a crucial role in peripheral nerve projection in mice. Neuron 1997; 19:995-1005.
62. Kawasaki T, Kitsukawa T, Bekku Y et al. A requirement for neuropilin-1 in embryonic vessel formation. Development 1999; 126:4895-4902.
63. Giger RJ, Cloutier JF, Sahay A et al. Neuropilin-2 is required in vivo for selective axon guidance responses to secreted semaphorins. Neuron 2000; 25:29-41.
64. Chen H, Bagri A, Zupicich JA et al. Neuropilin-2 regulates the development of selective cranial and sensory nerves and hippocampal mossy fiber projections. Neuron 2000; 25:43-56.
65. Takashima S, Kitakaze M, Asakura M et al. Targeting of both mouse neuropilin-1 and neuropilin-2 genes severely impairs developmental yolk sac and embryonic angiogenesis. Proc Natl Acad Sci USA 2002; 99:(in press).
66. Yamada Y, Takakura N, Yasue H et al. Exogenous clustered neuropilin 1 enhances vasculogenesis and angiogenesis. Blood 2001; 97:1671-1678.
67. Bagnard D, Vaillant C, Khuth ST et al. Semaphorin 3A-vascular endothelial growth factor-165 balance mediates migration and apoptosis of neural progenitor cells by the recruitment of shared receptor. J Neurosci 2001; 21:3332-3341.
68. Sondell M, Sundler F, Kanje M. Vascular endothelial growth factor is a neurotrophic factor which stimulates axonal outgrowth through the flk-1 receptor. Eur J Neurosci 2000; 12:4243-4254.
69. Jin KL, Mao XO, Greenberg DA. Vascular endothelial growth factor: direct neuroprotective effect in in vitro ischemia. Proc Natl Acad Sci USA 2000; 97:10242-10247.
70. Oosthuyse B, Moons L, Storkebaum E et al. Deletion of the hypoxia-response element in the vascular endothelial growth factor promoter causes motor neuron degeneration. Nat Genet 2001; 28:131-138.
71. Shirvan A, Shina R, Ziv I et al. Induction of neuronal apoptosis by Semaphorin3A-derived peptide. Brain Res Mol Brain Res 2000; 83:81-93.
72. Latil A, Bieche I, Pesche S et al. VEGF overexpression in clinically localized prostate tumors and neuropilin-1 overexpression in metastatic forms. Int J Cancer 2000; 89:167-171.
73. Lacal PM, Failla CM, Pagani E et al. Human melanoma cells secrete and respond to placenta growth factor and vascular endothelial growth factor. J Invest Dermatol 2000; 115:1000-1007.
74. Ding H, Wu X, Roncari L et al. Expression and regulation of neuropilin-1 in human astrocytomas. Int J Cancer 2000; 88:584-592.
75. Handa A, Tokunaga T, Tsuchida T et al. Neuropilin-2 expression affects the increased vascularization and is a prognostic factor in osteosarcoma. Int J Oncol 2000; 17:291-295.
76. Banerjee SK, Zoubine MN, Tran TM et al. Overexpression of vascular endothelial growth factor164 and its co-receptor neuropilin-1 in estrogen-induced rat pituitary tumors and GH3 rat pituitary tumor cells. Int J Oncol 2000; 16:253-260.
77. Miao HQ, Lee P, Lin H et al. Neuropilin-1 expression by tumor cells promotes tumor angiogenesis and progression. FASEB J 2000; 14:2532-9.
78. Benjamin LE, Keshet E. Conditional switching of vascular endothelial growth factor (VEGF) expression in tumors: induction of endothelial cell shedding and regression of hemangioblastoma-like vessels by VEGF withdrawal. Proc Natl Acad Sci USA 1997; 94:8761-8766.
79. Bachelder R, Crago A, Chung J et al. Vascular endothelial growth factor is an autocrine survival factor for neuropilin-expressing breast carcinoma cells. Cancer Res 2001; 61:5736-5740.
80. Xiang RH, Hensel CH, Garcia DK et al. Isolation of the human semaphorin III/F gene (SEMA3F) at chromosome 3p21, a region deleted in lung cancer. Genomics 1996; 32:39-48.
81. Roche J, Boldog F, Robinson M et al. Distinct 3p21.3 deletions in lung cancer and identification of a new human semaphorin. Oncogene 1996; 12:1289-1297.

82. Sekido Y, Bader S, Latif F et al. Human semaphorins A(V) and IV reside in the 3p21.3 small cell lung cancer deletion region and demonstrate distinct expression patterns. Proc Natl Acad Sci USA 1996; 93:4120-4125.
83. Brambilla E, Constantin B, Drabkin H et al. Semaphorin SEMA3F localization in malignant human lung and cell lines: A suggested role in cell adhesion and cell migration. Am J Pathol 2000; 156:939-950.
84. Tomizawa Y, Sekido Y, Kondo M et al. Inhibition of lung cancer cell growth and induction of apoptosis after reexpression of 3p21.3 candidate tumor suppressor gene SEMA3B. Proc Natl Acad Sci USA 2001; 98:13954-13959.
85. Tse C, Xiang RH, Bracht T et al. Human Semaphorin 3B (SEMA3B) Located at Chromosome 3p21.3 Suppresses Tumor Formation in an Adenocarcinoma Cell Line. Cancer Res 2002; 62:542-546.
86. Martin-Satue M, Blanco J. Identification of semaphorin E gene expression in metastatic human lung adenocarcinoma cells by mRNA differential display. J Surg Oncol Suppl 1999; 72:18-23.
87. Fujisawa H, Takagi S, Hirata T. Growth-associated expression of a membrane protein, neuropilin, in Xenopus optic nerve fibers. Dev Neurosci 1995; 17:343-349.
88. Zhang ZG, Tsang W, Zhang L et al. Up-regulation of neuropilin-1 in neovasculature after focal cerebral ischemia in the adult rat. J Cereb Blood Flow Metab 2001; 21:541-549.
89. Ishihama H, Ohbayashi M, Kurosawa N et al. Colocalization of neuropilin-1 and Flk-1 in retinal neovascularization in a mouse model of retinopathy. Invest Ophthalmol Vis Sci 2001; 42:1172-1178.
90. Ishida S, Shinoda K, Kawashima S et al. Coexpression of VEGF receptors VEGF-R2 and neuropilin-1 in proliferative diabetic retinopathy. Invest Ophthalmol Vis Sci 2000; 41:1649-1656.
91. Murohara T, Asahara T, Silver M et al. Nitric oxide synthase modulates angiogenesis in response to tissue ischemia. J Clin Invest 1998; 101:2567-2578.

NEUROPILIN-1 IN THE IMMUNE SYSTEM

Paul-Henri Romeo*, Valérie Lemarchandel and Rafaele Tordjman

SUMMARY

The neuropilin-1 (NRP1) and neuropilin-2 (NRP2) receptors can bind the class-3 semaphorin subfamily and the heparin-binding forms of vascular endothelial growth factor (VEGF) and placenta growth factor (PlGF). The functions of NRP1 and NRP2 have been extensively studied in neurons where they act in axon guidance and in endothelial cells where they promote angiogenesis and cell migration. In this chapter, we will present evidences indicating that neuropilin-1 is likely to mediate contacts between the dendritic cells and the T lymphocytes via homotypic interactions and is essential for the initiation of the primary immune response. These results emphasize the molecular similarities between the nervous and the immune systems and open new areas in the modulation of the immune response.

INTRODUCTION

The microorganisms that are encountered daily are detected and quickly destroyed by the cells involved in the innate immunity. However, if an infectious organism breaches this defense, an adaptive immune response, most often initiated by the dendritic cells, occurs. During this response, the dendritic cells efficiently capture, in the peripheral tissues, antigens from the microorganisms, process these antigens to form major histocompatibility complex molecule (MHC)-peptide complexes and migrate from the periphery to the T cell areas of the secondary lymphoid organs. Here, resting T cells encounter the antigen-carrying dendritic cells. This interaction creates an highly structured and localized adhesion complex called the immunological synapse at which specific ligands and costimulatory molecules trigger and sustain the T cell activation process (for reviews see refs. 1,2).

*Institut Cochin, Département Hématologie, INSERM U567, CNRS UMR 8104, Maternité Port-Royal, 123 Boulevard de Port-Royal, 75014 PARIS.

The formation of the immunological synapse is initiated by adhesive interactions between integrins such as LFA-1 and ICAM-1 or 2 or non-integrin molecules such as LFA-3 and CD2. These interactions overcome the barrier posed by the negatively charged glycocalyx of the dendritic and T cells, bring T cells and dendritic cells to within 15nm that is a distance allowing the T-cell antigen receptor (TCR) and the MHC-peptide complex interaction, and finally promote actin cytoskeleton rearrangements on dendritic and T cells. Then, the engaged TCRs are transported to the center of the immunological synapse while the engaged adhesion molecules are forced into a surrounding ring (Fig. 1). This large-scale molecular complex can be stable for several hours and sustained signaling on this time-scale is required for full T cell activation (for a review see ref. 3).

Although the list of receptors at the T cell surface that play a role in the establishment and maintenance of productive T cell-dendritic cell interactions seems endless, the dendritic cells receptors that mediate these initial interactions are not well known. Considering the analogy between dendritic cells and the neurons, originally described in 1868 by Paul Langherans,[4] and the role of neuropilin-1 in the regulation of the neurons cytoskeletal rearrangement we studied the expression and the role of this receptor in the primary immune response and this chapter will summarize the results we have obtained together with the prospects opened by these results.

NEUROPILIN-1 IS EXPRESSED BY DENDRITIC CELLS AND RESTING T CELLS

The differentiation of human monocytes into immature dendritic cells can be obtained in the presence of Granulocyte Macrophage-Colony Stimulating Factor (GM-CSF) and IL-4.[5] We found that the neuropilin-1 expression can be detected on immature dendritic cells but not monocytes. This cell-specific expression was found at the protein and the mRNA levels suggesting that the neuropilin-1 gene is transcriptionally activated during the dendritic cells differentiation. Further maturation of the dendritic cells in the presence of inflammatory cytokines (tumor necrosis factor-α or IL-1) did not significantly modified the neuropilin-1 expression. In vivo, neuropilin-1 expression was found in the dendritic cells present in human lymph nodes. Interestingly, T lymphocytes present in these lymph nodes were also stained and we found that resting T lymphocytes expressed neuropilin-1. Thus, neuropilin-1 is expressed by the two type of cells involved in the primary immune response.

T CELL-DENDRITIC CELL INTERACTION INDUCES NEUROPILIN-1 POLARIZATION IN T CELLS

During the formation of the immunological synapse, cell surface molecules such as CD3, CD4 or CD8 polarize on the effector T lymphocyte.[2] We showed that neuropilin-1 colocalizes with these proteins at the resting T cell-dendritic cell inter-

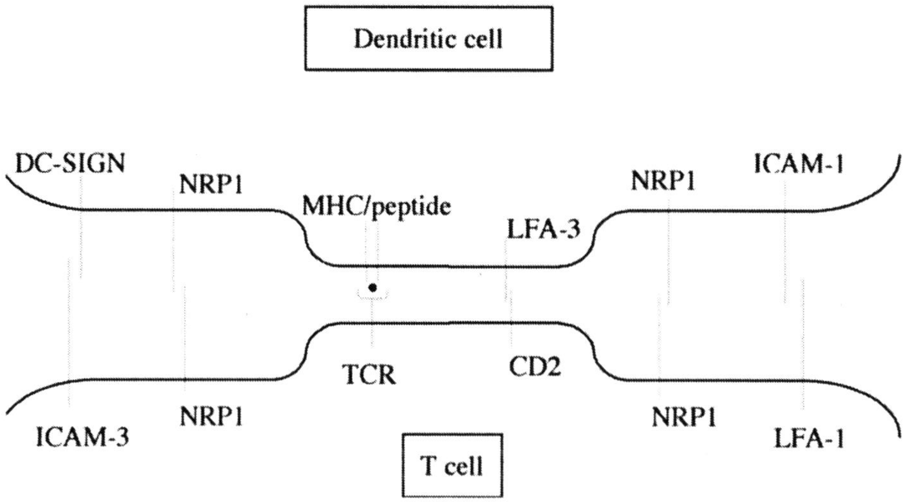

Figure 1. Dendritic and T cells interact through the formation of the immunological synapse. In this structure, the short interacting molecules (such as TCR/MHC peptide or CD2/LFA-3) are clustered within opposing membranes whereas long transmembrane proteins (such as ICAM-3/DC-SIGN, LFA-1/ICAM-1 and possibly NRP1) are excluded in a peripheral ring.

face indicating a polarization of neuropilin-1 towards the contact zone on T cells but not on the dendritic cells. We could not distinguish whether the increased intensity of neuropilin-1 immunofluorescence results from the redistribution of surface neuropilin-1 or is simply due to the close contact between the T and dendritic cells membranes which both expressed neuropilin-1 at the interface. However, as we found, in a few dendritic-T cell conjugates, a bipolar distribution of neuropilin-1 with neuropilin-1 concentrated at the DC-T cell contact zone and at the opposite pole of the T lymphocyte where no dendritic cell is present, we favor the redistribution of surface neuropilin-1 hypothesis.

NEUROPILIN-1 PROMOTES CELL-CELL INTERACTIONS

Although neuropilin-1 generally acts as a receptor for Sema 3A, VEGF or PlGF, it can also promote cell-cell contact through homophilic interactions in the absence of its ligands.[6] As mRNA expression of neuropilin-1 ligands was very low in the dendritic cells and the resting T cells, we studied whether neuropilin-1 can be involved in the dendritic-T cell contact through an homotypic interaction. Indeed, neuropilin-1 on resting T cells can mediate antigen-independent clustering of T cells on COS-7 cells engineered to express neuropilin-1 and blocking neuropilin-1 on resting T cells or allogeneic dendritic cells interferes with resting dendritic-T cell clustering. These results suggest that neuropilin-1 can act through homotypic interactions but do not rule out the presence of an uncharacterized semaphorin-like ligand that is expressed by resting T cells and dendritic cells.

NEUROPILIN-1 MEDIATES THE DENDRITIC CELLS-INDUCED PROLIFERATION OF RESTING T CELLS

To examine if these neuropilin-1-mediated interactions between the dendritic cells and the resting T cells are important in the initiation of primary immune responses, we studied the effects of blocking neuropilin-1 antibodies on the capacity of allogeneic dendritic cells to induce proliferation of resting T cells. The dendritic cells-induced proliferation of resting T cells was inhibited by 50% when T cells or dendritic cells were preincubated with blocking neuropilin antibodies but was not inhibited when a non blocking neuropilin antibody was used. This relative inhibition is equivalent to that obtained when antibodies against other proteins involved in the immunological synapse were used and reflects the numerous adhesion molecules involved in the dendritic-T cell contact. Taken together, these data argue that neuropilin-1-mediated interactions are necessary to initiate the primary immune response.

DISCUSSION

Neuropilin-1 is a multipurpose receptor. It can bind members of two non related families of ligands, the semaphorins (see Chapter 2) and the VEGF/PlGF (see Chapter 3, this book) or functions as a cell surface adhesion molecule through heterotypic interactions (see Chapter 1). It can also interact with at least four types of cell surface molecules: NGF receptor trkA,[7] VEGF receptors, plexins and L1-CAM (see Chapters 6, 7 and 8). Thus, it is not surprising to find neuropilin-1 as an essential component for very diverse biological functions. Indeed, neuropilin-1 has been shown to participate in the development of the nervous and cardiovascular systems and to regulate migration of neural crest cells and neural progenitors. In adult, neuropilin-1 seems to have an important role in intact and injured sensory neurons, in bone marrow stromal cells[8] and in angiogenesis. We now extend the field of action of neuropilin-1 as this receptor is essential for the initiation of the primary immune response.

The identification of neuropilin-1 as an additional mediator of antigen-independent naïve dendritic-T cell interaction will help the understanding of the role of the dendritic-T cell contact in the initiation of the primary immune response. This research will indeed take benefit from the results obtained on neural and endothelial cells but might also shed light on some unexpected results on neuropilin-1. Neuropilin-1 is known as a receptor for many different ligands and recently as a possible heterotypic partner of cell surface molecules such as L1-CAM.[9] We now present evidences that neuropilin-1 can also mediate cell-cell contact through homotypic interactions. These neuropilin/neuropilin interactions have already been

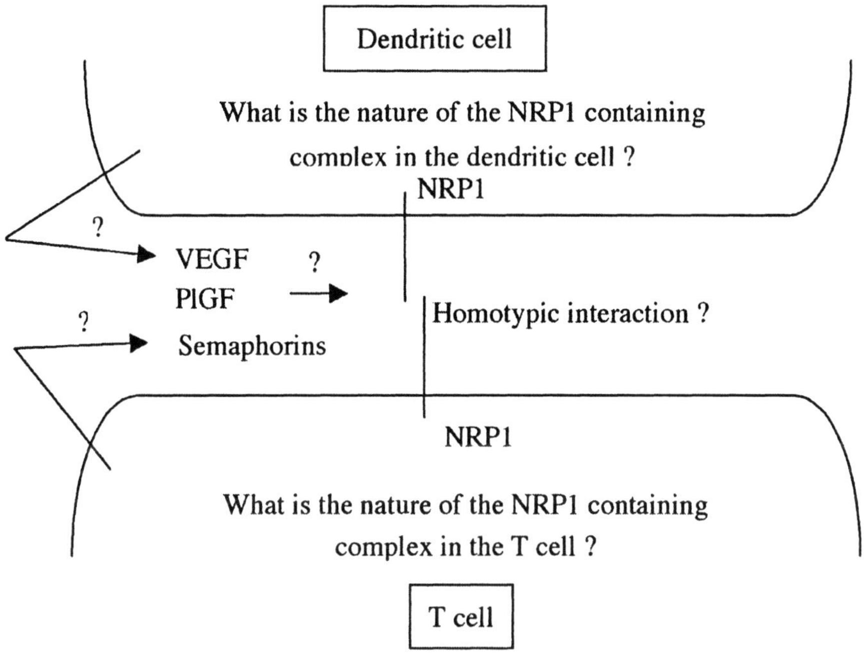

Figure 2. Further studies will determine (i) if an homotypic NRP1/NRP1 interaction occurs at the immunological synapse, (ii) the nature of the NRP1 containing complexes present in the dendritic and T cells and (iii) the expression of NRP1 ligands by the dendritic and T cells and their functions in the immune response.

demonstrated in the absence of ligand using transient transfections in COS cells. Our findings show the in vivo relevance of this observation. We could not rule out the presence of an unknown neuropilin-1 ligand on dendritic and T cells but the same inhibition obtained after incubation of dendritic and T cells or dendritic cells or T cells with the blocking neuropilin-1 antibody argues against the existence of this unknown neuropilin-1 ligand.

The functions of Semaphorins in both neuronal and non-neuronal cells are mediated by receptor complexes composed of members of the neuropilin and/or plexin protein families.[10] Ligand/receptor interaction results, in neuronal cells, in the axonal growth cone motility through cytoskeletal changes that are mediated by the Rho family GTPases.[11] As the dendritic cell cytoskeleton is critical for the formation of the immunological synapse,[12] neuropilin-1 might also act, albeit without ligand, as an inductor of similar dendritic cells cytoskeletal rearrangements. However, it remains to determine the nature of the neuropilin-1 partners in the dendritic cells and T cell and if these neuropilin-1 receptor complexes can transduce a signal in the dendritic or T cells (Fig. 2).

Finally, the neuropilin-1 ligands might also modulate the immune response. Activated T cells synthesize VEGF and we are currently studying the expression of all the NRP1 ligands in dendritic and T cells during the primary immune response. Assuming an homotypic NRP1/NRP1 interaction, these ligands might interfere with the NRP1/NRP1 contact and thus might be involved in the length of the dendritic/T cell contact. As this length is linked to the CD4 T cell polarization towards T helper 1 (T_H1) or T helper 2 (T_H2) cells,[13] neuropilin-1 and its ligands might become main players of the immune response and thus might become new important targets for treatment of diseases such as auto immune diseases or cancers where the immune response is disregulated.

REFERENCES

1. Banchereau J, Steinman RM. Dendritic cells and the control of immunity. Nature 1998; 392:245-252.
2. Dustin ML, Cooper JA. The immunological synapse and the actin cytoskeleton: Molecular hardware for T cell signaling. Nature Immunol 2000; 1:23-29.
3. Lanzavecchia A, Sallusto F. Antigen decoding by T lymphocytes: From synapses to fate determination. Nature Immunol 2001; 2:487-492.
4. Langerhans P. Uber die nerven der menschlichen haut. Virchows Arch Path Anat 1868; 44:325-337.
5. Geissmann F, Prost C, Monnet JP et al. Transforming growth factor beta1, in the presence of granulocyte/macrophage colony-stimulating factor and interleukin 4, induces differentiation of human peripheral blood monocytes into dendritic Langerhans cells. J Exp Med 1998; 187:961-966.
6. Chen H, He Z, Bagri A et al. Semaphorin-neuropilin interactions underlying sympathetic axon responses to class III semaphorins. Neuron 1998; 21:1283-1290
7. Tuttle R, Yano H, Chao MV et al. Neuropilin-1 and trk exist in a complex regulated by NGF. Soc Neurosci 2000; Abst 26:579.
8. Tordjman R, Ortega N, Coulombel L et al. Neuropilin-1 is expressed on bone marrow stromal cells: a novel interaction with hematopoietic cells? Blood 1999; 94:2301-2309
9. Castellani V, Chedotal A, Schachner M et al. Analysis of the L1-deficient mouse phenotype reveals cross-talk between Sema3A and L1 signaling pathways in axonal guidance. Neuron 2000; 27:237-249.
10. Takahashi T, Fournier A, Nakamura F et al. Plexin-neuropilin-1 complexes form functional semaphorin-3A receptors. Cell 1999; 99:59-69.
11. Driessens MH, Hu H, Nobes CD et al. Plexin-B semaphorin receptors interact directly with active Rac and regulate the actin cytoskeleton by activating Rho Curr Biol 2001; 11:339-344.
12. Al-Alwan MM, Rowden G, Lee TD et al. The dendritic cell cytoskeleton is critical for the formation of the immunological synapse. J Immunol 2001; 166:1452-1456
13. Iezzi G, Scheidegger D, Lanzavecchia A. Migration and function of antigen-primed nonpolarized T lymphocytes in vivo. J Exp Med 2001; 193:987-993

STRUCTURAL AND FUNCTIONAL RELATION OF NEUROPILINS

Fumio Nakamura[1] and Yoshio Goshima[1,2]

SUMMARY

Neuropilin is a type I transmembrane protein and the molecular mass is 120 kDa. Two homologues, Neuropilin-1 and -2, are identified. The primary structure of Neuropilin-1 and Neuropilin-2 is well conserved and is divided into four domains, CUB (a1/a2) domain, FV/FVIII (b1/b2) domain, MAM (c) domain, and (d) domain that contains a transmembrane and a short cytoplasmic region. Both Neuropilin-1 and Neuropilin-2 have truncated and secreted form of splice variants. Neuropilins act as a receptor for two different extracellular ligands, class 3 semaphorins and specific isoforms of vascular endothelial growth factor. In both cases, neuropilin requires an additional transmembrane molecule to exhibit biological activity. Plexin-A is essential for class 3 semaphorin signaling. Vascular endothelial cell growth factor (VEGF) receptor is the major receptor for VEGF and neuropilin acts as isoform specific co-receptor for VEGF. The CUB and FV/FVIII domains of Neuropilin are the binding sites of semaphorin and VEGF. The MAM domain mediates semaphorin signaling to Plexin-A. Cross talk between semaphorin and VEGF on neuropilin suggests that class 3 semaphorins and the secreted forms of neuropilin act as antagonists to VEGF and its related growth factors.

INTRODUCTION

Neuropilin (NRP) is a single-spanning membrane protein and the molecular mass is 120 kDa. The protein has been firstly identified as an antigen of a specific antibody A5, which recognized the developmental stage of *Xenopus* optic nerve.[1]

[1]Department of Molecular Pharmacology and Neurobiology, Yokohama City University School of Medicine, 3-9 Fukuura, Kanazawa-ku, Yokohama, Kanagawa, 236-0004, JAPAN and [2]CREST, Japan Science and Technology Corporation(JST), Japan.

The protein is unique to vertebrates, and in so far, zebra fish, frog, chick, mouse, rat and human NRP homologues have been partially or completely identified. No homologous proteins in invertebrates have been reported. The expression pattern of NRP varies among species. Based on the expression pattern and stage in chick[2] and mice,[3] it has been speculated that NRP is involved in the formation of nervous system. However, the role of NRP had not been exactly revealed until two ligands were identified.

In 1997, two groups independently reported that NRP is the receptor for Sema3A, one of the class 3 semaphorin.[4,5] Since an orthologue of NRP was reported as Neuropilin-2 (NRP2) at that time, NRP was renamed as Neuropilin-1 (NRP1). Sema3A is one of the members of Semaphorin family that is involved in the axon guidance during embryonic developmental stages.[6] The application of anti-NRP1 antibody blocked Sema3A-induced growth cone collapse of rat E14 dorsal root ganglion (DRG) cells, confirming that the protein acted as a functional receptor for Sema3A. This fact is further strengthened by mutant mouse studies. DRG growth cones of $NRP1^{-/-}$ mice did not respond to Sema3A.[7] Both $Sema3A^{-/-}$ and $NRP1^{-/-}$ mutant mice exhibited similar phenotype in nervous systems, such as aberrant and defasciculated peripheral nerve projection.[7,8]

In 1998, different aspect of NRP1 was revealed. NRP1 also acts as an isoform specific receptor for Vascular Endothelial cell Growth Factor (VEGF).[9] VEGF is a growth factor that stimulates the migration and proliferation of endothelial cells.[10,11] One of the major isoforms of VEGF, $VEGF_{165}$, binds to NRP1. The biological role of NRP1 in vascular system is proved by the studies of $NRP1^{-/-}$ mutant mice and NRP1 transgenic mice.[12,13] Both $NRP1^{-/-}$ and transgenic mice died before birth and the vascular regression in the $NRP1^{-/-}$ embryos was in marked contrast to the overproduction of vessels in the embryos transgenic NRP1.

A unique character of NRP is that the protein is unable to generate the intracellular signaling. Instead, additional molecules are required to exhibit the biological function of Semaphorin and VEGF. For semaphorin signaling, Plexin-A (Plex-A) acts as an essential signal transducer.[14] For VEGF signaling, VEGFR1(flt-1), or VEGFR2 (KDR/flk-1) are the major and functional receptor molecule and NRP serves as modulator.[9]

This section describes the structure of NRP1 and NRP2 at genome and protein level, then discusses the relation of the structure and the biological function of NRPs.

PRIMARY STRUCTURE AND GENOMIC STRUCTURE OF NEUROPILIN

Both NRP1 and NRP2 have transmembrane and truncated forms.[15] The transmembrane forms of NRP1 and NRP2 share similar primary structure (Fig. 1). Following a short stretch of secretion signal, NRP1 and NRP2 consist of four different domains, two repeats of CUB domain (a1/a2), two repeats of FV/VIII domain (b1/b2), a MAM (c) domain, and a fourth domain (d) that contains transmembrane and

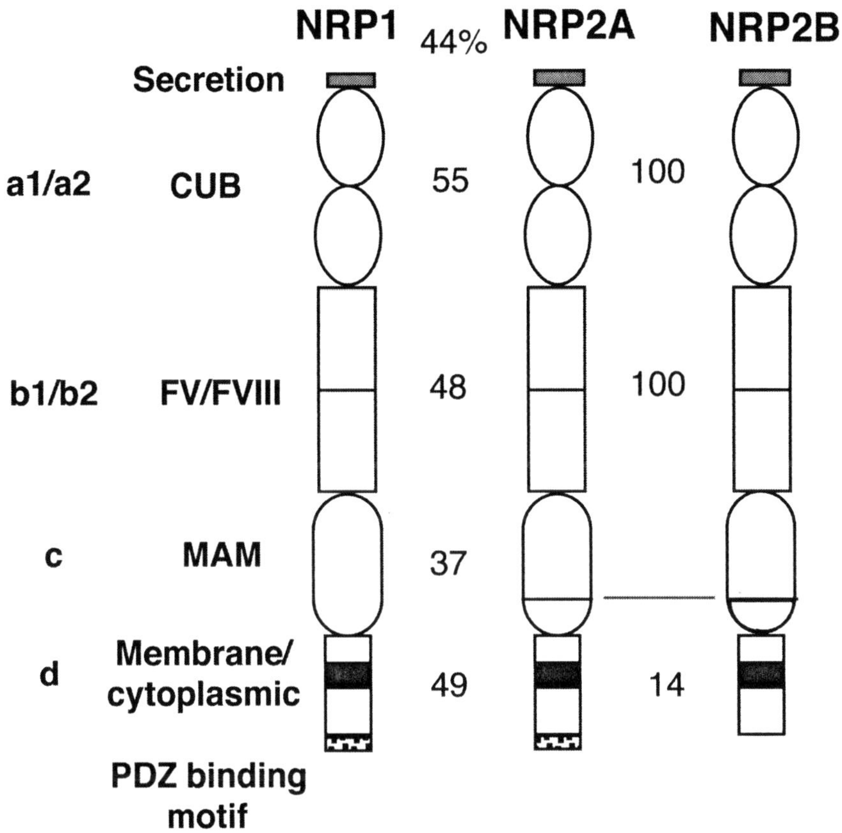

Figure 1. Primary structure of Neuropilins
Two homologues of NRPs, NRP1 and NRP2, are shown. NRP consists of four domains, two repeats of complement (CUB, a1/a2) domain, two repeats of coagulation factor(FV/FVIII, b1/b2) domain, a MAM (Meprin, A5, and Receptor protein-tyrosine phosphatase μ) (c) domain, and a transmembrane region (d). The amino acid identity of each domain is shown in the middle of NRP1 and NRP2. NRP1 and NRP2A possess a PDZ binding motif at the carboxyl termini that interact with NIP.

relatively short 40 to 43 amino acid cytoplasmic region. The first CUB domains have significant homology with complement factor C1s/C1r, Bone Morphogenetic Protein 1(BMP1), and Tolloid proteins. The second FV/VIII domain shares the homology with coagulation factor FV/VIII, one of the receptor type tyrosine kinase DDR, and discoidin-1. The third domain MAM is the abbreviation of meprin, A5 (former name of NRP), and receptor protein-tyrosine phosphatase mu and kappa (PTPμ, κ).

Full length NRP2 has two major alternate splice variants, NRP2A and NRP2B.[16] While NRP1 and NRP2A show 44% amino acid identity in entire regions, the fourth domain of NRP2B is unrelated to NRP1 or NRP2A. The amino acid identity of each domain, CUB, FV/FVIII, and MAM is 45%, 48% and 35%, respectively. In the fourth domain, NRP1 and NRP2A shares 49% identity, while NRP1 and NRP2B shares only 15% identity. Although the fourth domain has no apparent homology with other proteins, NRP1 and NRP2A possess a PSD-95/DIg/ZO-1 (PDZ) binding motif at the carboxyl termini. It has been shown that a Neuropilin-1 Interacting Protein (NIP) is associated with the NRP1 carboxyl terminus.[17]

The completion of human genome sequence[18] and a report from Klagsbrun laboratory[15] prompted us to examine the genomic structure of NRPs.(Fig. 2) The loci of human NRP1 and NRP2 genes are 10p12 and 2q34, respectively. The loci of mouse versions are chromosome 8 and 1. Human NRP1 gene spans a length of approximately 157 kb and is split into 19 exons. Human NRP2 is 115 kb length and is divided into 23 exons. The exon and intron numbers in this section are according to the assignment of the Human Genome Sequence.[18] Some of the exons within the contig are not used for encoding NRPs. These exons are shown as white half size boxes in Fig 2. The exons 3 and 9 in NRP1 locus generate different mRNAs. NRP2 locus contains three exons (5, 6, 21) unrelated to encode NRP2 mRNA. The relation between these exotic exons and NRP genes is currently unknown.

The location of exon-intron junction is similar for these two genes. Fourteen of the 16 splice sites are conserved between NRP1 and NRP2 gene. These sites are found in the exons encoding CUB and FV/FVIII domains, correlating well with the amino acid homologies of these domains. The splicing points corresponding to the regions of MAM and transmembrane are less conserved. The similar structure of NRP1 and NRP2 suggests a duplication of these genes in the evolution of vertebrates.

Alternative splices variants of full length NRP2, NRP2A and NRP2B, are generated by the splicing of the exons 19, 20, and 22. The fourth domain of NRP2A and NRP2B is encoded by exon 22 and exon 20, respectively. Comparing NRP2A and NRP2B, NRP2A is 17 amino acid length longer than NRP2B between the junction point and transmembrane region. Another alternate splicing in exon 19 generates 5 amino acid insertion flanked with the sequence encoded by exon 20 or exon 22. Then four splice variants, NRP2A17, NRP2A22, NRP2B0, and NRP2B5 are generated. In mice, two additional variants, NRP2A0 and NRP2A5, were also reported.[16] NRP2A and NRP2B show slightly different expression patterns in adult humans.[15] NRP2A is predominantly expressed in liver, placenta, lung, intestine, heart and kidney, while NRP2B is found in heart and skeletal muscle. Significant level of expression of both forms is detectable in adult brain. The role of structural difference between NRP2A and NRP2B is not clear. However, since the carboxyl termini of NRP2B variants are not homologous to the termini of NRP1 and NRP2A, NRP2B may not bind to NIP.

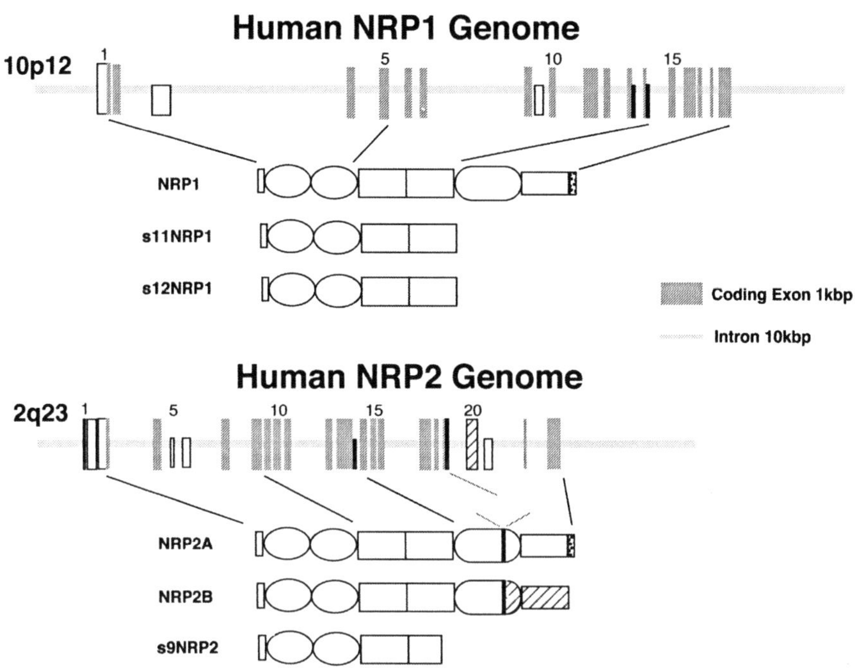

Figure 2. Genomic structure of human NRP1 and NRP2
The genome structure of human NRP1 and NRP2 is shown. The loci of human NRP1 and NRP2 are 10p12 and 2q34, respectively. The boxes indicate exons predicted from the comparison with reported mRNAs. Exon numbers are indicated above the boxes. Shaded boxes correspond to the coding sequence of NRP1 or NRP2. The exons indicated as white half size boxes (exons 3 and 9 in NRP1, exons 5, 6, and 21 in NRP2) are not found in NRP mRNAs. Two black half sized boxes in NRP1 gene and one in NRP2 gene represent the intron-derived insertion of soluble NRPs. A full size black box in NRP2 is alternatively spliced 15 bases in exon 19. The region inserts 5 amino acid (GENFK) to both NRP2A and NRP2B. A diagonal lined box exon 20 encodes the fourth domain of NRP2B.

It has been identified two truncated forms of NRP1, s11NRP1 and s12NRP1, and one short form of NRP2, s9NRP2.[15,19] All of the truncated forms are generated by the use of alternate polyadenylation signals in the specific introns. Within the mRNA of s11NRP1, the sequence of exon 13 is flanked with a 1866 base intron 13-derived sequence encoding a 84 unique amino acid sequence. The mRNA of s12NRP1 has a 28 base intron 14-derived sequence after the exon 14 junction. Since exon 13 and 14 encode FV/FVIII domain, s11NRP1 and s12NRP1 consist of only CUB and FV/FVIII domains. No apparent hydrophobic regions are encoded by the intron-derived sequences. Then, two truncated forms of NRP1 are secreted proteins. These variants are predominantly expressed in placenta, liver, heart, kidney and lung. The expression of both forms in brain is lower than other tissues.

A soluble form of NRP2, s9NRP2, is also generated by the same manner as the truncated forms of NRP1.[15] The mRNA of s9NRP2 is flanked with 144 bp intron

13-derived sequence after exon 13. The 144-bp sequence contains a stop codon and a polyadenylation site. Then s9NRP2 consists of two CUB domains, the first b1 of FV/FVIII domain, a part of b2 domain, and a 8 unique amino acid sequence encoded by the intron 9.

BINDING PROPERTIES OF NRP DOMAINS

Semaphorins are a large family of transmembrane and secreted proteins are mainly involved in axon guidance.[6] All of the semaphorins possess a 550 amino acid Semaphorin (Sema) domain at their amino termini and are divided into 8 classes based on species, amino acid sequence, and structural similarity. Classes 1 and 2 are invertebrate semaphorins, classes 3 to 7 are vertebrates, and class V is viral semaphorins. Classes 1, 4, 5, 6 and 7 are membrane bound proteins, while classes 2, 3 and V are secreted semaphorins.

In so far, at least three different types of binding proteins were reported as semaphorin receptors, Neuropilin,[4,5] Plexin,[20] and CD72.[21] Classes 1 and 2 invertebrate semaphorins bind to Plexin.[22] In vertebrates, class 4, 5, 6, 7 semaphorins bind to Plexin B, C or A directly, and class V semaphorins bind to Plexin C.[23] Plexins possess a Sema domain at the amino termini, suggesting that the proteins are distant ancestors of semaphorins. It is interesting to note that CD72 serves as a functional receptor for Sema4D (CD100) in lymphocytes.[21] CD72 is a 45 kDa type II membrane protein that belongs to the C-type lectin. Sema4D binds to CD72 with a Kd of 300 nM and augments the effect of CD40 on B cell responses, such as proliferation.

A unique feature of class 3 semaphorins is that they bind to NRPs but not to plexins. Six members of semaphorins, 3A, 3B, 3C, 3D, 3E and 3F, belong to class 3. This class of secreted protein consists of a Sema domain, one immunoglobulin domain and a basic amino acid rich region. Class 3 Semaphorins form a homodimer through a disulfide bond in the basic rich region, which is critical to exhibit biological activity.[24]

NRP1 and NRP2 exhibits different specificity to class 3 semaphorins. NRP1 binds to all class 3 semaphorins while NRP2 binds to Sema3B, Sema3C, Sema3D, Sema3E, and Sema3F but not to Sema3A.[25] Although NRP1 and NRP2 display relatively large spectrum of binding to class 3 semaphorins, this does not account for the specific effect of each member of class 3 semaphorins. For example, the DRG express only NRP1 and the growth cones are sensitive to Sema3A but not Sema3B or 3C, while sympathetic ganglion cells that express both NRP1 and NRP2 are repelled by Sema3A, Sema3B, and Sema3C.[26]

The characterization of class 3 semaphorin binding to NRP1 and NRP2 has been conducted on Sema3A, 3C and 3F. He and Tessier-Lavigne showed that Sema3A binds to NRP1 with two different regions, the Sema domain and the basic rich region.[4] The broad binding specificity of NRP1 reflects the binding of the basic rich region of class 3 semaphorins. For example, full length Sema3F binds to NRP1 and NRP2, while Sema-Ig domain of Sema3F shows higher affinity to NRP2. In contrast, the basic rich region of Sema3F binds effectively NRP1 but not NRP2.[27] The discrepancy of binding specificity between Sema-Ig domain and basic rich region compromises broad spectrum of class 3 semaphorin binding to NRPs.

However, the binding of Sema domain to NRPs, but not of basic rich region, is the initial step to elicit the biological function of class 3 semaphorins. The growth cone collapsing activity is retained in a chimera protein consisting of the Sema domain of Sema3A fused to the Fc region of human IgG1.[28] The importance of Sema domain is also supported by the study of chimera of chick Sema3A (Collapsin-1) and Sema3D (Collapsin-2),[29] in which Sema3A but not Sema3D induced growth cone collapse of embryonic chick DRG cells. In this study, swapping of 70 amino acid region of Sema3A and Sema3D within the Sema domain altered the growth cone collapse activity of both proteins. Each member of class 3 semaphorin uses different combinations of NRP1 and NRP2 to exhibit repulsive action. For instance, Sema3A action is mediated by NRP1 whereas Sema3F is mediated by NRP2. On the other hand, the action of Sema3C requires both NRP1 and NRP2.[27]

Five different groups performed deletion and chimera analyses on NRP1 and NRP2.[27,30-33] Nakamura et al[30] and Chen et al[27] demonstrated that the CUB domains of NRPs are the binding site for Sema domain of class 3 semaphorins. A deletion mutant NRP1Δ276-797 that contains only the CUB domain of NRP1 was able to bind the Sema-Ig portion of Sema3A (Fig. 3). The CUB domain also determines the binding preference of NRP2. A NRP1/NRP2 chimera 2111 in which CUB domain of NRP1 was substituted with the one of NRP2 bound to Sema3C but not to Sema3A.[30] The selective binding of 2111 chimera to the semaphorins was similar to the specificity of NRP2. These results indicate that the CUB domain is the primary binding site of the Sema domain. In contrast, the basic rich region of Sema3A was bound to the boundary of CUB and FV/FVIII domains. The basic rich region bound to NRP1Δ18-253 (Fig. 3)[31] but not to NRP1Δ18-282.[30] This suggests that the region including a 27 amino acid stretch from 255 to 282 of NRP1 is critical for the binding of the basic rich region. It has been shown that the FV/FVIII domain of NRP1 is involved in the binding of Vascular Endothelial cell Growth Factor (VEGF)[31] and in NRP1-mediated cell adhesion.[33]

While the CUB and FV/FVIII domains are involved in the binding of class 3 semaphorin and other ligands, the MAM domain participates in the signal transmission of semaphorins. The MAM domain of NRP shares homology with receptor protein tyrosine phosphatase μ and meprin. It has been shown that the MAM domain participates in the oligomerization of NRPs.[30]

The functional role of the MAM domain in Class 3 semaphorin signaling was demonstrated by Sema3A responsiveness of the chick retinal ganglion cells engineered to express full length NRP1 or a series of NRP1 deletion mutant. Chick retinal ganglion cells lack normally NRP1 expression, therefore the growth cones do not respond to Sema3A. Herpes Simplex Virus vector mediated expression of NRP1 in these cells renders Sema3A responsiveness. Using this system, series of deletion and chimera mutants of NRP1 were introduced and examined.[30] The experiments showed that the MAM domain deleted mutant was unable to transmit Sema3A signaling. This was consistent with other studies: a MAM domain deleted

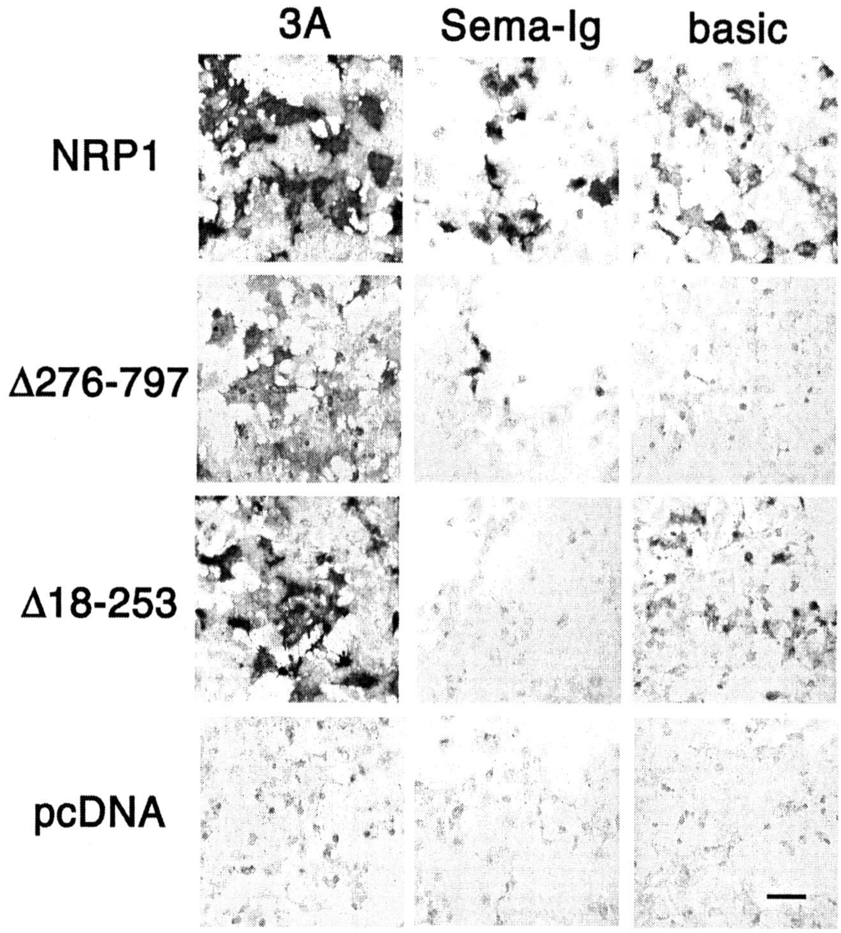

Figure 3. The binding of Sema3A to NRP1 mutants
Either NRP1, a NRP1 mutant deleted FV/FVIII and MAM domains (NRP1Δ276-797), or a partial deletion of CUB domain (NRP1Δ18-253) was expressed in COS-7 cells. Sema3A and its deletion mutants were fused with human placental alkaline phosphatase (AP). The cells were incubated with AP-Sema3A (300pM), Sema-Ig-AP (4nM), or AP-basic rich region (1nM) for 60 min. Bound AP fusion proteins were visualized. NRP1Δ276-797 bound the Sema-Ig but not the basic rich region, whereas NRP1Δ18-253 bound the basic rich region, suggesting the CUB domain is the binding site for the Sema domain of Sema3A. Scale bar, 100 μm.

NRP1 mutant also showed a dominant-negative effect on Sema3A-induced growth cone collapse of embryonic chick sympathetic neurons.[32] The antibody directed against the MAM domain of NRP2 blocked the Sema3F-induced growth cone collapse of sympathetic ganglion cells.[31] These results indicates that the MAM domain is the interface of semaphorin signaling. Then, what kind of molecule interacts with the MAM domain and acts as a signal transducer?

NEUROPILIN-1 INTERACTING PROTEIN BINDS TO THE CARBOXYL TERMINUS OF NRP1

Shortly after the identification of NRP1 as a receptor for Sema3A, efforts to identify the downstream of Sema3A signaling began. A straightforward approach was the use of Yeast two hybrid system for searching binding partners of the intracellular region of NRP1. This method successfully isolated Neuropilin-1 Interacting Protein (NIP), which recognizes the four amino acids of carboxyl terminus of NRP1 (Figs. 1 and 4).[17] NIP is a 40kDa protein and possesses one central domain that shares significant homology with PSD-95/Dig/ZO-1(PDZ). It has been shown that this domain binds to the four amino acid sequence of NRP1 carboxyl terminus, Tyr-Ser-Glu-Ala. While NRP2A also contains a similar sequence Cys-Ser-Glu-Ala at the carboxyl terminus, the carboxyl terminus of NRP2B is different from NRP1 and NRP2A. The deletion of three amino acids from the carboxyl terminus of NRP1 diminishes the interaction with NIP.

NIP has been independently cloned as RGS-GAIP-interacting protein (GIPC)[34] and SEMCAP-1.[35] RGS-GAIP is a GTPase-activating protein (GAP) for Gαi subunits, which is localized on clathrin-coated vesicles. The carboxyl terminus of RGS-GAIP is Ser-Glu-Ala, which exactly matches to NRP1. SEMCAP-1 was identified as an interacting protein with the intracellular region of Sema4C (M-SemaF). The carboxyl terminus of Sema4C is Glu-Ser-Ser-Val, limiting homology to the termini of NRP1 and RGS-GAIP. Then, X-Ser-X-Ala/Val is the consensus motif of NIP/GAIP/SEMCAP-1 binding site. Since NIP is a cytoplasmic protein and the extracellular domain of NRP transmits the Semaphorin signal to another protein, the involvement of NIP in class 3 semaphorin signaling has not been well defined.

ADDITIONAL RECEPTOR REQUIRED FOR SIGNAL TRANSDUCTION

As described in Chapter 5, plexin is the receptor for certain class of semaphorins. *Drosophila* Sema1a interacts with *Drosophila* Plex-A.[22] Class 7 semaphorins and viral Semaphorins bind to Plex-C.[20,23] Class 3 semaphorins do not bind plexins directly, however, Plexin-A forms complex with NRP and acts as signaling molecule.[14] This is supported by the following facts. While COS-7 cells expressing NRP1 bound Sema3A with a Kd of 1 nM, Plex-A1 co-expression increased the affinity of NRP1 to Sema3A about five-fold with a Kd of 0.2 nM. Furthermore, COS-7 cells co-expressing NRP1 and Plex-A1 presented contracted cell morphology after Sema3A exposure. This phenomenon resembles to the collapsed state of growth cones. Interestingly, the cells expressing MAM domain deleted NRP1 and Plex-A1 did not exhibit high affinity binding site for Sema3A nor alter the

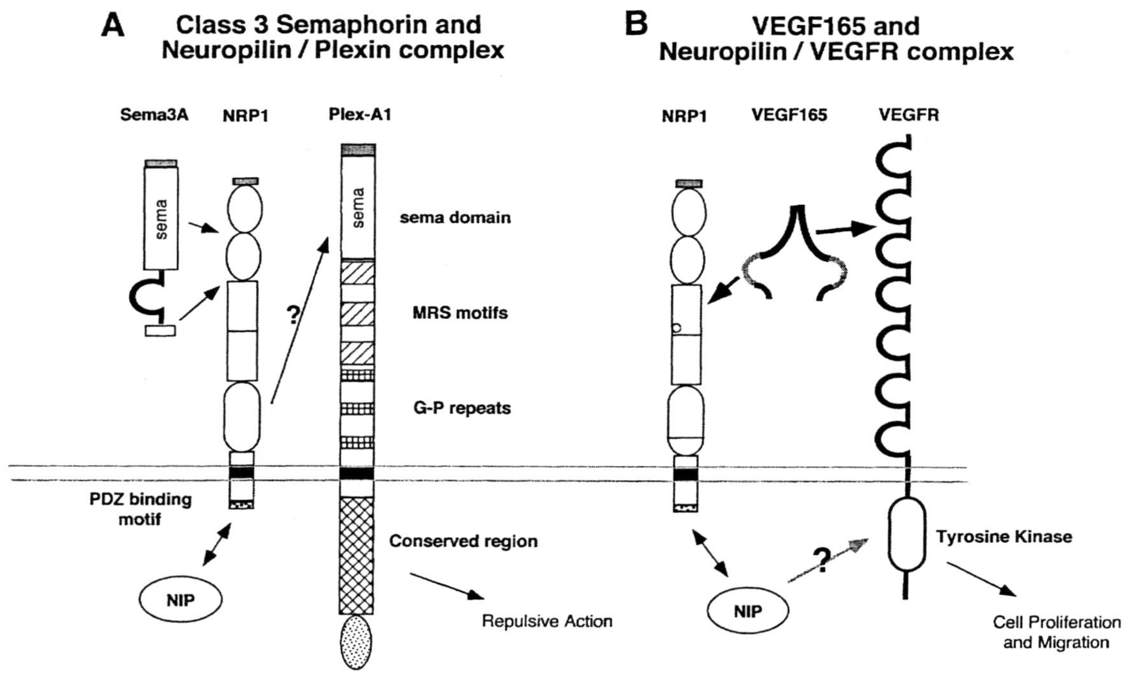

Figure 4. Schematic representation of Neuropilin and its interacting proteins
Two extracellular ligands bind to NRP. A) Class 3 semaphorins bind to the CUB and FV/FVIII domains of NRP. Plexin-A associates with NRP and transmits signals inside the cells. NIP binds to the carboxyl termini of NRP1 and NRP2A. The protein also interacts with the carboxyl termini of Sema4F and RGS-GAIP. The protein may facilitate the oligomerization of NRP and/or other NIP binding proteins. B) A VEGF isoform containing exon 7 derived 44 amino acid sequence, $VEGF_{165}$, binds to the FV/FVIII domain of NRP1. The region encoded by the exons from 1 to 5 of VEGF-A gene binds directly to VEGFR. VEGF dimerizes VEGFR, then activates the tyrosine kinase in cytoplasmic region. The oligomerization of NRP1 or NRP2A by NIP may be required for $VEGF_{165}$ binding. (See text for details)

morphology after Sema3A stimulation. This fact clearly demonstrates that Sema3A-collapsing signal is transmitted from MAM domain to Plex-A1. Finally, the overexpression of a mutant Plex-A1 without intra-cytoplasmic domain blocked Sema3A-induced growth cone collapse of chick DRG. This indicates that the Plex-A1 mutant acts as dominant negative manner and the cytoplasmic tail of Plex-A1 initiates an intracellular signal of collapse response. Other members of Plexin-A, Plex-A2 and Plex-A3 have been shown to mediate class 3 semaphorin signals.[36,37]

Vascular endothelial cell growth factor (VEGF-A) is a potent factor that induces the formation of blood vessels.[10] VEGF forms a 40-45 K homodimer and has low homology with platelet-derived growth factor. Five different polypeptides, 121, 145, 165, 189 and 206 amino acids ($VEGF_{121}$-$VEGF_{206}$) are generated by alternative splicing from VEGF-A gene, each of them capable of making an active homodimer. $VEGF_{121}$ and $VEGF_{165}$ are the most abundant forms. A unique 44 amino acid sequence of $VEGF_{165}$ is derived from exon 7. $VEGF_{121}$ possesses the biological activity of VEGF, the stimulation of proliferation and migration of endothelial cells. Comparing to $VEGF_{121}$, $VEGF_{165}$ is a more potent mitogen for endothelial cells, suggesting the modulating role of the unique sequence of $VEGF_{165}$.

At least two receptor type tyrosine kinases, VEGFR1 (Flt-1) and VEGFR2 (KDR/Flk-1), serve as VEGF receptors (Fig. 4).[10] Both VEGFR1 and VEGFR2 have 7 immunoglobulin repeats in the extracellular region, a transmembrane domain, and an intracellular tyrosine kinase domain. The biological activity of VEGF is exhibited through the dimerization of the receptor, subsequently leading to the activation of the tyrosine kinase.

Besides these main receptors, Soker et al[9] reported that NRP1 is a specific receptor for $VEGF_{165}$ but not for $VEGF_{121}$. This finding is quite consistent with the expression pattern of NRP1 as well as mutant NRP1 mice phenotype.[12] The interactions between VEGFR and NRP are fully detailed in Chapter 7.

Porcine aortic endothelial (PAE) cells lack expression of NRP1 and VEGFR2, allowing to the expression and functional study of these receptors in vascular cells. When NRP1 was expressed in PAE cells, $VEGF_{165}$ bound to NRP1 with a Kd of 0.3 nM. While VEGF stimulated the migration of PAE cells expressing VEGFR2, $VEGF_{165}$ could not alter the migration of PAE cells expressing NRP1. However, co-expression of NRP1 and VEGFR2 in PAE cells augmented the migration upon the stimulation of $VEGF_{165}$, comparing to single-expression of VEGFR2. The direct interaction of NRP1 and VEGFR2 was also shown by co-immunoprecipitation.[38] These results demonstrate that NRP1 acts as a co-receptor of $VEGF_{165}$. NRP2 also serves as co-receptor for $VEGF_{145}$ as well as for $VEGF_{165}$. Other related homologues of VEGF, placenta growth factor-2 (PlGF-2) have been shown to bind NRP1 and NRP2.

Although the function of NIP in Class 3 semaphorin signaling has not been demonstrated, NIP may participate in the signal transduction of VEGF. One hint has been provided by the study of the signal transduction of $VEGF_{165}$ through NRP2 and VEGFR1 in PAE cells.[39] While wild type NRP2A could bind $VEGF_{165}$, a tagged NRP2A containing a myc epitope at the carboxyl terminus, which disrupted the

PDZ binding motif, could not bind to $VEGF_{165}$. In contrast, Sema3F binding to NRP2A was not affected by the addition of myc tag. Considering the fact that NIP is a broadly expressed protein[17, 35] and it probably binds to the Ser-Glu-Ala motif of NRP2A carboxyl terminus, the homo- or hetero-oligomerization of NRP2A through NIP may be required for the binding of $VEGF_{165}$.

It is of interest whether two distinct ligands may interfere with each other on one receptor molecule. The study using transient expression of NRP1 and VEGFR2 in COS-7 cells demonstrated that Sema3A inhibits the binding of $VEGF_{165}$ to NRP1. Sema3A also antagonizes the $VEGF_{165}$-induced migration of PAE cells co-expressing NRP1 and VEGFR2. Indeed, Semaphorin-NRP interaction plays positive or negative regulatory role in lung branching morphogenesis,[40] and Sema3B and Sema3F have been implicated as tumor suppressor genes in human lung small cell carcinoma.[41] These semaphorins may suppress the expansion of tumors by antagonizing VEGF and its related growth factors.

Bagnard et al[42] reported that migration and apoptosis of neural progenitor cells was regulated by the balance of Sema3A and $VEGF_{165}$. Furthermore, they observed that Sema3A activates the tyrosine kinase of VEGFR1. This suggests that VEGFR1 may serve as an additional component of semaphorin receptor, at least during the migration stage of neural progenitor. Further investigation is required to prove this idea.

As mentioned earlier, soluble forms of NRP1 and NRP2 are predominantly expressed in non-neuronal tissues. These truncated forms of NRP may also act as an inhibitor of VEGF-induced vascular formation. When rat prostate carcinoma-derived cell lines were injected to a rat host, tumor masses were formed in various organs. These masses were invaded by numerous blood vessels because of the production of VEGF. When the same cell lines expressing soluble s12NRP1 were injected, most of the malignant cells in the mass were destined to apoptotic degradation.[19] Poor formation of blood vessels in the tumors was also observed. In this case, the s12NRP1 probably antagonized VEGF action to malignant cells and to invading vascular endothelial cells. This opens the possibility of the therapeutic use of soluble NRPs as anti-tumor reagents.

CONCLUDING REMARKS

Three distinct extracellular domains of NRP play important roles in semaphorin and VEGF signaling (Fig. 4). The CUB and FV/FVIII domains serve as the binding sites for the two ligands. The MAM domain acts as a signaling interface to Plex-A, at least in the case of class 3 semaphorin signal transduction (Fig. 4A). VEGF165 binds to both NRP and VEGFR (Fig. 4B). A PDZ protein, NIP binds to the carboxyl termini of NRP1 and NRP2A. NIP may play important role for VEGF signaling. Recent accumulating findings begin to resolve the mysterious action of class 3 semaphorins as tumor suppressor genes. Class 3 semaphorins seem to antagonize the binding of VEGF or VEGF related growth factors to NRPs. Then the Semaphorins regulate precisely the strength of these factors and maintain appropriate growth of various organs. The soluble forms of NRPs may also play a similar role by quenching the growth factors.

Now the role of NRPs as a multiple ligand receptor emerges. Furthermore, NRPs require an appropriate transmembrane molecule in accordance with the bound ligand to exhibit the biological function. As the signal transducer, class 3 semaphorins use Plex-A. L1, another cell surface molecule, is also thought to interact with NRP1 and to be involved in Sema3A signaling.[43] In the case of VEGF, NRP acts as co-receptor of VEGFR. All of the molecules described in this Chapter, NRPs, Plex-A, L1, and VEGFR are single membrane spanning proteins and structurally unrelated or distant. This suggests that some of transmembrane proteins may form functional hetero-oligomers with other unrelated proteins rather than homo-oligomerization as seen in the activation of receptor tyrosine kinases. Verifying the known single membrane spanning proteins from this aspect may find alternate new role of those proteins.

ACKNOWLEDGMENTS

We thank to Professor Stephen M. Strittmatter at Yale University for providing AP-Sema3A, Sema-Ig-AP, AP-basic, NRP1 and NRP1Δ276-797(1001) expression vectors.

REFERENCES

1. Takagi S, Hirata T, Agata K et al. The A5 antigen, a candidate for the neuronal recognition molecule, has homologies to complement components and coagulation factors. Neuron 1991; 7(2):295-307.
2. Takagi S, Kasuya Y, Shimizu M et al. Expression of a cell adhesion molecule, neuropilin, in the developing chick nervous system. Dev Biol 1995; 170(1):207-222.
3. Kawakami A, Kitsukawa T, Takagi S et al. Developmentally regulated expression of a cell surface protein, neuropilin, in the mouse nervous system. J Neurobiol 1996; 29(1):1-17.
4. He Z, Tessier-Lavigne M. Neuropilin is a receptor for the axonal chemorepellent Semaphorin III. Cell 1997; 90(4):739-751.
5. Kolodkin AL, Levengood DV, Rowe EG et al. Neuropilin is a semaphorin III receptor. Cell 1997; 90(4):753-762.
6. Semaphorin Nomenclature Committee. Unified nomenclature for the semaphorins/collapsins. Semaphorin Nomenclature Committee. Cell 1999; 97(5):551-552.
7. Kitsukawa T, Shimizu M, Sanbo M et al. Neuropilin-semaphorin III/D-mediated chemorepulsive signals play a crucial role in peripheral nerve projection in mice. Neuron 1997; 19(5):995-1005.
8. Taniguchi M, Yuasa S, Fujisawa H et al. Disruption of semaphorin III/D gene causes severe abnormality in peripheral nerve projection. Neuron 1997; 19(3):519-530.
9. Soker S, Takashima S, Miao HQ et al. Neuropilin-1 is expressed by endothelial and tumor cells as an isoform-specific receptor for vascular endothelial growth factor. Cell 1998; 92(6):735-745.
10. Klagsbrun M, D'Amore PA. Vascular endothelial growth factor and its receptors. Cytokine Growth Factor Rev 1996; 7(3):259-270.
11. Risau W. Mechanisms of angiogenesis. Nature 1997; 386(6626):671-674.
12. Kawasaki T, Kitsukawa T, Bekku Y et al. A requirement for neuropilin-1 in embryonic vessel formation. Development 1999; 126(21):4895-4902.

13. Kitsukawa T, Shimono A, Kawakami A et al. Overexpression of a membrane protein, neuropilin, in chimeric mice causes anomalies in the cardiovascular system, nervous system and limbs. Development 1995; 121(12):4309-4318.
14. Takahashi T, Fournier A, Nakamura F et al. Plexin-neuropilin-1 complexes form functional semaphorin-3A receptors. Cell 1999; 99(1):59-69.
15. Rossignol M, Gagnon ML, Klagsbrun M. Genomic organization of human neuropilin-1 and neuropilin-2 genes: identification and distribution of splice variants and soluble isoforms. Genomics 2000; 70(2):211-222.
16. Chen H, Chedotal A, He Z et al. Neuropilin-2, a novel member of the neuropilin family, is a high affinity receptor for the semaphorins Sema E and Sema IV but not Sema III. Neuron 1997; 19(3):547-559.
17. Cai H, Reed RR. Cloning and characterization of neuropilin-1-interacting protein: a PSD-95/Dlg/ZO-1 domain-containing protein that interacts with the cytoplasmic domain of neuropilin-1. J Neurosci 1999; 19(15):6519-6527.
18. International Human Genome Sequencing Consortium. Initial sequencing and analysis of the human genome. Nature 2001; 409(6822):860-921.
19. Gagnon ML, Bielenberg DR, Gechtman Z et al. Identification of a natural soluble neuropilin-1 that binds vascular endothelial growth factor: In vivo expression and antitumor activity. Proc Natl Acad Sci USA 2000; 97(6):2573-2578.
20. Tamagnone L, Artigiani S, Chen H et al. Plexins are a large family of receptors for trans-membrane, secreted, and GPI-anchored semaphorins in vertebrates. Cell 1999; 99(1):71-80.
21. Kumanogoh A, Watanabe C, Lee I et al. Identification of CD72 as a lymphocyte receptor for the class IV semaphorin CD100: A novel mechanism for regulating B cell signaling. Immunity 2000; 13(5):621-631.
22. Winberg ML, Noordermeer JN, Tamagnone L et al. Plexin A is a neuronal semaphorin receptor that controls axon guidance. Cell 1998; 95(7):903-916.
23. Comeau MR, Johnson R, DuBose RF et al. A poxvirus-encoded semaphorin induces cytokine production from monocytes and binds to a novel cellular semaphorin receptor, VESPR. Immunity 1998; 8(4):473-482.
24. Klostermann A, Lohrum M, Adams RH et al. The chemorepulsive activity of the axonal guidance signal semaphorin D requires dimerization. J Biol Chem 1998; 273(13):7326-7331.
25. Nakamura F, Kalb RG, Strittmatter SM. Molecular basis of semaphorin-mediated axon guidance. J Neurobiol 2000; 44(2):219-229.
26. Takahashi T, Nakamura F, Jin Z et al. Semaphorins A and E act as antagonists of neuropilin-1 and agonists of neuropilin-2 receptors. Nat Neurosci 1998; 1(6):487-493.
27. Chen H, He Z, Bagri A et al. Semaphorin-neuropilin interactions underlying sympathetic axon responses to class III semaphorins. Neuron 1998; 21(6):1283-1290.
28. Eickholt BJ, Morrow R, Walsh FS et al. Structural features of collapsin required for biological activity and distribution of binding sites in the developing chick. Mol Cell Neurosci 1997; 9(5-6):358-371.
29. Feiner L, Koppel AM, Kobayashi H et al. Secreted chick semaphorins bind recombinant neuropilin with similar affinities but bind different subsets of neurons in situ. Neuron 1997; 19(3):539-545.
30. Nakamura F, Tanaka M, Takahashi T et al. Neuropilin-1 extracellular domains mediate semaphorin D/III-induced growth cone collapse. Neuron 1998; 21(5):1093-1100.
31. Giger RJ, Urquhart ER, Gillespie SK et al. Neuropilin-2 is a receptor for semaphorin IV: Insight into the structural basis of receptor function and specificity. Neuron 1998; 21(5):1079-1092.
32. Renzi MJ, Feiner L, Koppel AM et al. A dominant negative receptor for specific secreted semaphorins is generated by deleting an extracellular domain from neuropilin-1. J Neurosci 1999; 19(18):7870-7880.
33. Shimizu M, Murakami Y, Suto F et al. Determination of cell adhesion sites of neuropilin-1. J Cell Biol 2000; 148(6):1283-1293.

34. De Vries L, Lou X, Zhao G et al. A PDZ domain containing protein, interacts specifically with the C terminus of RGS-GAIP. Proc Natl Acad Sci USA 1998; 95(21):12340-12345.
35. Wang LH, Kalb RG, Strittmatter SM. A PDZ protein regulates the distribution of the transmembrane semaphorin, M-SemF. J Biol Chem 1999; 274(20):14137-14146.
36. Takahashi T, Strittmatter SM. PlexinA1 autoinhibition by the plexin sema domain. Neuron 2001; 29(2):429-439.
37. Cheng HJ, Bagri A, Yaron A et al. Plexin-a3 mediates semaphorin signaling and regulates the development of hippocampal axonal projections. Neuron 2001; 32(2):249-263.
38. Whitaker GB, Limberg BJ, Rosenbaum JS. Vascular endothelial growth factor receptor-2 and neuropilin-1 form a receptor complex that is responsible for the differential signaling potency of VEGF(165) and VEGF(121). J Biol Chem 2001; 276(27):25520-25531.
39. Gluzman-Poltorak Z, Cohen T, Shibuya M et al. Vascular endothelial growth factor receptor-1 and neuropilin-2 form complexes. J Biol Chem 2001; 276(22):18688-18694.
40. Kagoshima M, Ito T, Kitamura H et al. Diverse gene expression and function of semaphorins in developing lung: positive and negative regulatory roles of semaphorins in lung branching morphogenesis. Genes Cells 2001; 6(6):559-571.
41. Sekido Y, Bader S, Latif F et al. Human semaphorins A(V) and IV reside in the 3p21.3 small cell lung cancer deletion region and demonstrate distinct expression patterns. Proc Natl Acad Sci USA 1996; 93(9):4120-4125.
42. Bagnard D, Vaillant C, Khuth ST et al. Semaphorin 3A-vascular endothelial growth factor-165 balance mediates migration and apoptosis of neural progenitor cells by the recruitment of shared receptor. J Neurosci 2001; 21(10):3332-3341.
43. Castellani V, Chedotal A, Schachner M et al. Analysis of the L1-deficient mouse phenotype reveals cross-talk between Sema3A and L1 signaling pathways in axonal guidance. Neuron 2000; 27(2):237-249.

THE FUNCTION OF NEUROPILIN/PLEXIN COMPLEXES

Andreas W. Püschel

SUMMARY

Neuropilins bind the secreted class 3 semaphorins with high affinity but require a member of the plexin family to form receptors that are able to activate downstream signal transduction cascades. In this receptor complex neuropilins act as the ligand-binding subunit while plexins function as the signal-transducing subunit in the induction of cytoskeletal collapse by semaphorins. The cytoplasmic domain is highly conserved within the plexin family and interacts with Rho-like GTPases.

INTRODUCTION

Relatively soon after the identification of neuropilin-1 (NRP1) as an essential component of the Sema3A receptor[1] it became apparent that the two members of the neuropilin family, NRP1 and NRP2, are not sufficient to form functional and specific receptors for class 3 semaphorins on their own. Embryonic day (E8) chick retinal ganglion neurons that do not bind and respond to Sema3A[2,3] and do not express NRP1 become susceptible to the repulsive effects of Sema3A upon expression of NRP1 from viral vectors.[4] This assay allowed Nakamura et al[4] to show that the cytoplasmic domain of NRP1 is dispensable for its ability to confer Sema3A-sensitivity to retinal axons. Replacement of the cytoplasmic and transmembrane domains of NRP1 by a heterologous sequence or a GPI-anchor did not impair its ability to confer Sema3A-sensitivity. As deletion of its cytoplasmic domain did not affect the ability of NRP1 to act as a Sema3A receptor additional

Institut für Allgemeine Zoologie und Genetik, Westfälische Wilhelms-Universität, Schlossplatz 5, D-48149 Münster Germany.

receptor subunit(s), present in E8 chick retinal ganglion neurons, must be responsible for activating downstream signal transduction cascades.

NEUROPILINS FORM THE LIGAND-BINDING SUBUNIT OF THE SEMA3A RECEPTOR

Semaphorins exert very specific effects on NRP1 and -2 expressing neurons. Sema3A has strong repulsive effects on NRP1-positive neurites such as sensory or sympathetic axons, while Sema3C or 3F affect only sympathetic axons that also express NRP2 but do not repel sensory axons.[3,5-8] Binding assays and the effects of blocking antibodies suggested that NRP1 acts as receptor specific for Sema3A and NRP2 for Sema3F while NRP1/NRP2 heterodimers are required for repulsion by Sema3C.[7,9] Additional studies showed, however, that NRP1 and -2 probably bind all secreted class 3 semaphorins and differ only in their affinity for individual family members (Rohm and Püschel, unpublished results).[10,11] Thus, the differential binding of class 3 semaphorins to NRP1 and -2 cannot fully account for their selective effects on axons. Takahashi et al[12] reported that Sema3B and 3C act as agonists for receptors containing NRP2 whereas they behave as competitive antagonists for Sema3A on receptors containing NRP1. The specific effects of secreted class 3 semaphorins may, therefore, result from differences in their ability to activate NRP1 containing receptors. The molecular determinants for agonistic or antagonistic effects of semaphorins remain to be identified, however.

Neuropilins differ not only in their specificity but also their affinity from semaphorin receptors present on neurons. The affinity of NRP1 for Sema3A (dissociation constant (K_D) = 0.33 - 1.15 nM)[1,9] is at least one order of magnitude lower than that determined for neuronal Sema3A-binding sites (0.03 nM) or the EC_{50} for the collapse of sympathetic (0.05 nM) by Sema3A.[13,14] Therefore, additional receptor components may be required not only to form signaling-competent receptors but also high affinity binding sites specific for a single class 3 semaphorin.

PLEXINS ACT AS THE SIGNAL-TRANSDUCING SUBUNIT OF SEMAPHORIN RECEPTORS

With the identification of VESPR (Plexin-C1) and *Drosophila* Plexin-A as receptors for the vaccinia virus encoded semaphorin A39R (SemaVa) and Drosophila Sema1a, respectively, the plexins emerged as candidates for the missing signal-transducing subunit.[15,16] Indeed, it was shown independently by three groups that several plexins (Plexin-A1, -A2, -A3, and -B1) could interact with both NRP1 and NRP2. Both neuropilins formed complexes with plexins independently of the presence of ligands.[17-19] The plexins are a family of large integral membrane proteins with a highly conserved cytoplasmic domain. At their amino-terminus they contain a semaphorin domain which shows a moderate degree of sequence identity to the corresponding domain of semaphorins that includes 14 conserved cystein residues.[16] Together with the receptor protein tyrosine kinases Met, Ron, and Sea, plexins and semaphorins form a superfamily of semaphorin domain-containing proteins.[18,20,21] In addition, their extracellular domains are characterized by two or three Met-related sequence (MRS) repeats also found in many other proteins.[18,22]

Plexins are widely expressed in the developing central and peripheral nervous system including hippocampal, cortical, sensory, and sympathetic neurons.[23,24] mRNAs of all four A-type plexins can be detected in dorsal root ganglia where Plexin-A1 shows the lowest and Plexin-A3 and -A4 the highest expression levels.

Plexins do not directly bind class 3 semaphorins but, in a complex with neuropilins, are essential for mediating the repulsive effects of Sema3A.[17-19] Deletion of the conserved cytoplasmic domain of Plexin-A1 or -A2 results in a dominant-negative receptor that can suppress repulsion by Sema3A in Xenopus motor neurons and mouse sensory neurons.[18,19] Co-expression of NRP1 and Plexin-A1 in COS-7 cells allows the reconstitution of a functional Sema3A receptor in a heterologous system.[17] These results demonstrate that the Sema3A receptor consists of NRP1 as the ligand binding subunit and a member of the A-type plexins as the signal-transducing subunit. It remains to be investigated if plexins are also involved in mediating the attractive effects of class 3 semaphorins.[25,26]

PLEXINS ARE ESSENTIAL COMPONENTS OF THE SEMA3A RECEPTOR

In addition to its role in signal transduction, complex formation with plexins also changes the ligand-binding properties of neuropilins. Neuropilin/plexin complexes display an increased specificity for secreted semaphorins.[17,19] A NRP1/Plexin-A1 complex prefers Sema3A over Sema3C or Sema3F while NRP2/Plexin-A2 preferentially binds Sema3F. The biochemical basis for the increased specificity remains, however, controversial. While Takahashi et al[17] report an increase in the affinity of NRP1/Plexin-A1 for Sema3A, Rohm et al[19] demonstrated an increase in the number of Sema3A binding sites detectable on transfected 293T cells when NRP1 and Plexin-A1 were co-expressed in comparison to cells expressing only NRP1.[17,19] In contrast, the number of binding sites for Sema3C was reduced. The reason for this discrepancy is presently unclear but may result from differences in the assay systems.

The genomes of *Drosophila melanogaster* and *Caenorhabditis elegans* both contain two plexin genes.[16,18] In mammals at least 9 plexins were identified[18,24] that can be subdivided into 4 classes (Fig. 1). Plexin-B1 and Plexin-C1 were identified as receptors for the membrane-bound semaphorins Sema4D and Sema7A, respectively.[18] A-type plexins can act as receptor subunits for secreted class 3 semaphorins. At present, it is not clear if all A-type plexins are able to function as the signal transducing subunit of the Sema3A receptor. Experiments in COS-7 cells suggest that Plexin-A1 and -A2 are able to act as Sema3A receptors while Plexin-A3 is not.[17] In contrast, inactivation of the mouse Plexin-A3 gene convincingly shows

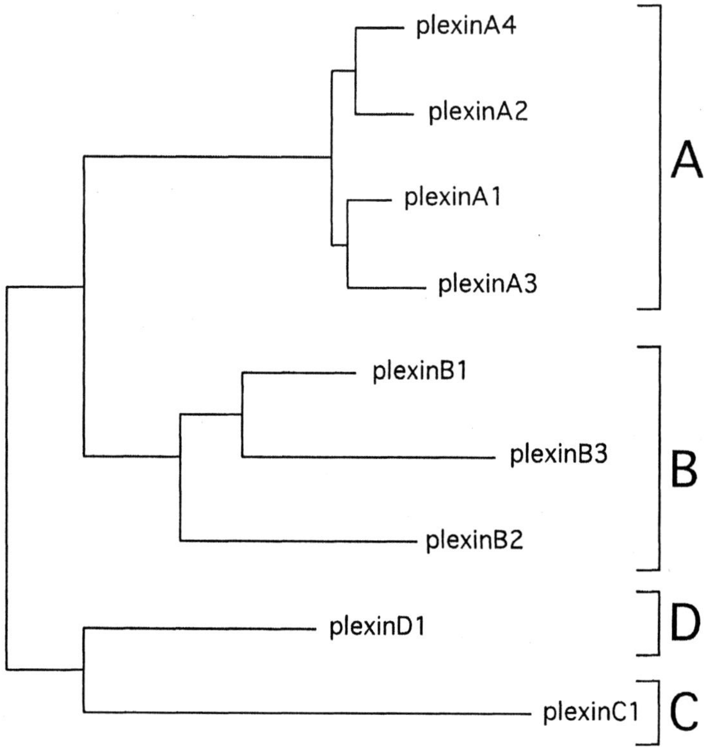

Figure 1. Plexins form a large gene family. Sequence comparison of cytoplasmic domains using the programs CLUSTAL and PILEUP (HUSAR 3.0 software; dkfz, Heidelberg) allows to distinguish 4 subgroups of plexins (A, B, C, and D).

that Plexin-A3 can transduce repulsive signals and contribute to Sema3A and Sema3F signaling in vivo.[24] While axons from explanted Plexin-A3$^{-/-}$ dorsal root and superior cervical ganglia showed only a reduced response to Sema3A, the sensitivity of SCG axons to Sema3F was completely abolished. Similar observations were made for hippocampal axons. Homozygous Plexin-A3 mutant mice were viable and fertile and showed only minor defects in peripheral innervation. The ophtalmic branch of the trigeminal nerve was defasciculated in E10.5 to E12.5 mice. In addition, defects in hippocampal projections were observed. The discrepancy between genetic approaches and in vitro assays may indicate limitations of the COS-7 collapse assay. The inability of Plexin-A3 to mediate cell collapse in response to Sema3A in COS-7 cells may, alternatively, suggest inefficient post-translational processing or trafficking of Plexin-A3. Indeed, we observed that Plexin-A3 is retained to a large extent in intracellular compartments in COS-7 cells (Rohm and Püschel, unpublished results).

In addition to the binding of neuropilins, intramolecular interactions of plexins are involved in determining their signaling properties as semaphorin receptors.[27] The amino-terminal half of Plexin-A1 contains a semaphorin domain that associates with the carboxy-terminal half of the plexin ectodomain and thereby keeps Plexin-A1 in the inactive state. Consequently, deletion of the semaphorin domain or the complete ectodomain of Plexin-A1 results in the formation of a receptor that is constitutively active both in a heterologous cell system (COS-7 cells) and in neurons. The semaphorin domain and the C-terminal half of the Plexin-A1 ectodomain interact independently with NRP1. In NRP1 the binding site for the semaphorin domain of Plexin-A1 does not overlap with that for Sema3A as excess Sema3A does not prevent interaction of NRP1 with the Plexin-A1 semaphorin domain. Both remain associated with NRP1 after addition of Sema3A. These results suggest that prior to ligand binding the intramolecular interaction between the subdomains of Plexin-A1 results in a self-inhibition which is released upon binding of Sema3A to NRP1.

THE ROLE OF GTPASES FOR SIGNAL TRANSDUCTION BY PLEXINS

The high sequence conservation of their cytoplasmic domains suggests an involvement of plexins in activating downstream signaling cascades upon activation by semaphorins. This conclusion was confirmed by the observation that a deletion of this domain not only prevents plexin-induced collapse in COS-7 cells and neurons but also results in the formation of a dominant-negatively acting receptor.[17-19] The cytoplasmic domain does not contain any homology to the catalytic domains of other well characterized receptors. A more detailed analysis, however, reveals sequence similarities to GTPase activating proteins (GAPs) specific for Ras-like GTPases (Fig. 2).[28] GTPases act as molecular switches that regulate multiple cellular processes by activating downstream effectors when in the GTP-bound form. GAPs stimulate their intrinsic GTPase activity and terminate signaling by GTPases. The GAP homology is split into two blocks separated by a sequence of variable size in different plexins that is less well conserved between different family members. The GAP homologies include two arginine residues that correspond to the essential catalytic residues found in rasGAPs. These arginines are essential also for the function of plexins as semaphorin receptors and mutation of either amino acid residue suffices to completely block the ability of Plexin-A1 to induce the collapse of COS-7 cells.[28]

Their homology to GAPs suggests that plexins might regulate the activity of or interact with small monomeric GTPases of the Ras or Rho families. Indeed, several groups reported a differential binding of GTPases to plexins.[28-32] While Rac interacts with Plexin-B1, Plexin-A1 binds the Rho-like GTPases Rnd1 and RhoD, and *Drosophila* Plexin-B forms complexes with Rho and Rac. Ligand binding increased the interaction of Plexin-B1 and Rac1. Rnd1 and RhoD have antagonistic effects on

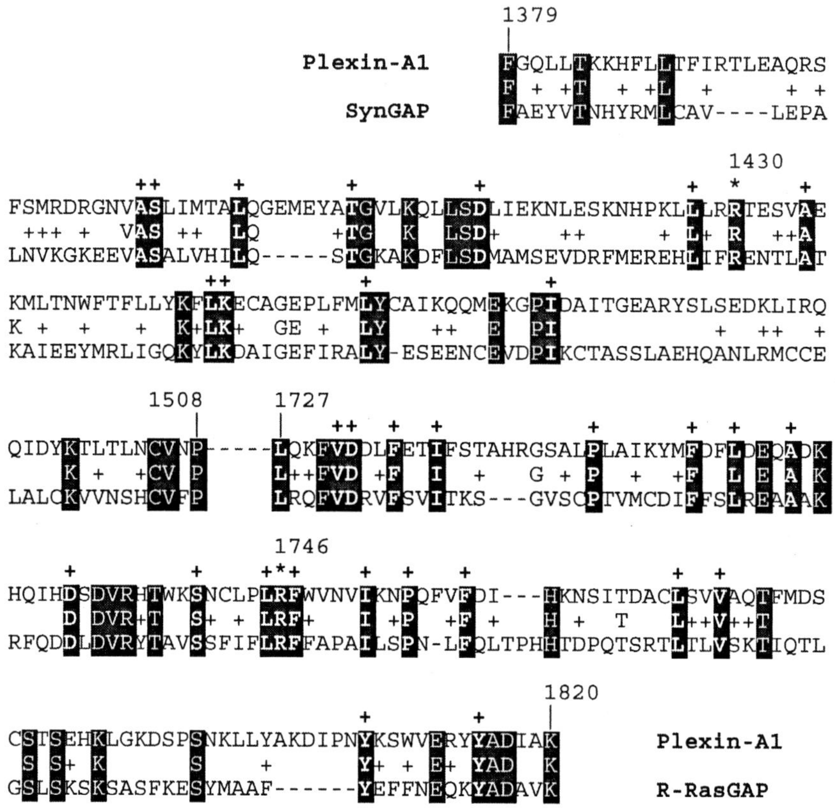

Figure 2. The cytoplasmic domain of Plexin-A1 shows sequence similarity to RasGAPs. Alignments of the amino acid sequence of Plexin-A1 with partial sequences of SynGAP or R-RasGAP are shown. Residues conserved between all Plexins and ras GAPs are indicated by + signs and the conserved arginine residues R1430 and R1746 by asterisks above the Plexin-A1 sequence.

the activity of Plexin-A1 and probably are involved in the initiation and not the execution of cytoskeletal collapse by Plexin-A1.[32] They appear to act upstream of Plexin-A1 to regulate its activity as a Sema3A receptor. Whereas interaction of Rnd1 and Plexin-A1 triggers signaling by Plexin-A1 and results in cytoskeletal collapse in the absence of any ligand, binding of RhoD has the opposite effect and blocks Plexin-A1 activity (Fig. 3). Activation of Plexin-A1 by Rnd1 may be a prerequisite for its ability to induce cell or growth cone collapse upon Sema3A binding. The regulation of Plexin-A1 activity by Rnd1 and RhoD does not require the presence of NRP1. The role of NRP1 in the receptor complex, thus, may be restricted largely to ligand-binding.

Work from many labs demonstrated that Rho-like GTPases are central regulators of cytoskeletal dynamics that control the organization of actin filaments and

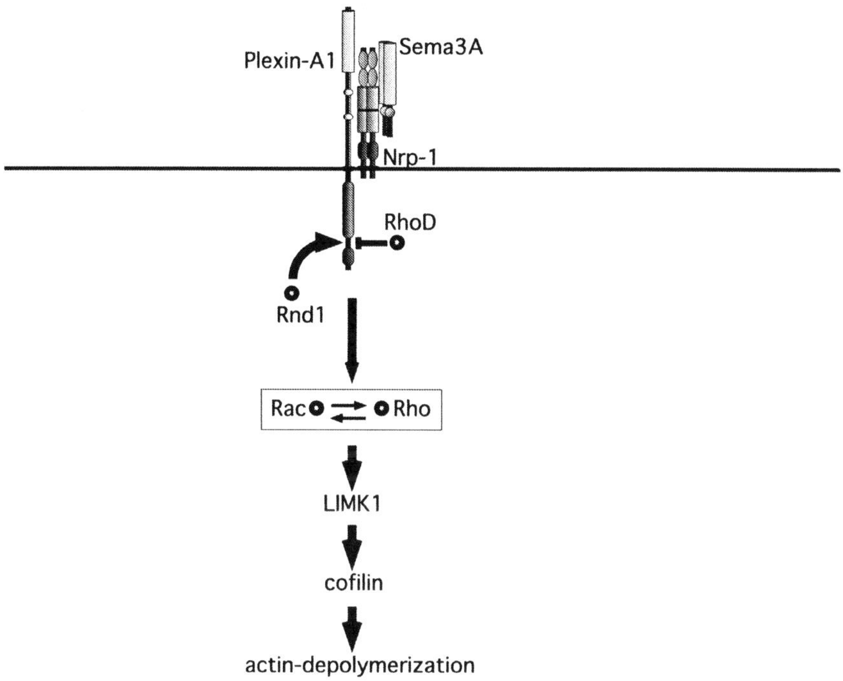

Figure 3. Regulation of Plexin-A1 activity by RhoD and Rnd1. The Rho-like GTPases Rnd1 and RhoD interact with Plexin-A1 and regulate its activity. Interaction of Plexin-A1 and Rnd1 results in an activation of Plexin-A1 and downstream signaling events that probably shift the balance of Rac and Rho activity towards actin depolymerization. This process is blocked by interaction of Plexin-A1 with RhoD. The molecular components that link Rac and Rho to active Plexin-A1 are presently unknown.

microtubuli.[33] Rho and Rac activity determines the cellular morphology of fibroblasts and neurons. Activation of Rho induces neurite retraction while active Rac promotes it.[34-38] Therefore, it is not surprising that the Rho-like GTPase Rac1 is also involved in mediating actin depolymerization during Sema3A-induced growth cone collapse. Inhibition of Rac activity by introducing dominant-negative RacN17 blocks Sema3A-induced growth cone collapse which suggests that the Sema3A receptor regulates the activity of Rho-like GTPases.[39-41] Downstream of Rac, phosphorylation of cofilin, a regulator of actin polymerization, by LIM Kinase 1 is essential for Sema3A induced growth cone collapse.[42] Most of the signaling events, however, that translate the binding of Sema3A to its receptor into changes in the balance of Rho and Rac activity and structural changes of the cytoskeleton remain to be elucidated.

OPEN QUESTIONS

Despite tremendous progress in understanding the function of neuropilins many questions remain to be addressed. The formation of a NRP1/Plexin-A1 complex is essential for the function of the Sema3A receptor. The question remains, however, if additional subunits are required to form receptors specific for a single semaphorin. Both neuropilins interact not only with A-type plexins but also with at least one B-type plexin (Plexin-B1) that does not require NRP1 to act as a receptor for Sema4D. Can plexins other than A-type plexins mediate effects of the secreted semaphorins? Do distinct neuropilin/plexin complexes differ in their properties? Finally, neuropilins interact not only with plexins but also with at least two other proteins, the cell adhesion molecule L1 and the receptor for vascular endothelial growth factor VEGFR1.[43-45] It has not been investigated so far if neuropilin/plexin complexes are still able to interact with these proteins or if they form mutually exclusive complexes.

REFERENCES

1. He Z, Tessier-Lavigne M. Neuropilin is a receptor for the axonal chemorepellent Semaphorin III. Cell 1997; 90:739-751.
2. Takagi S, Kasuya Y, Shimizu M et al. Expression of a cell adhesion molecule, neuropilin, in the developing chick nervous system. Dev Biol 1995; 170:207-222.
3. Luo Y, Raible D, Raper JA. Collapsin: a protein in brain that induces the collapse and paralysis of neuronal growth cones. Cell 1993; 75:217-227.
4. Nakamura F, Tanaka M, Takahashi T et al. Neuropilin-1 extracellular domains mediate semaphorin D/III-induced growth cone collapse. Neuron 1998; 21:1093-1100.
5. Messersmith EK, Leonardo ED, Shatz CJ et al. Semaphorin III can function as a selective chemorepellent to pattern sensory projections in the spinal cord. Neuron 1995; 14:949-959.
6. Püschel AW, Adams RH, Betz H. Murine semaphorin D/collapsin is a member of a diverse gene family and creates domains inhibitory for axonal extension. Neuron 1995; 14:941-948.
7. Giger RJ, Urquhart ER, Gillespie SK et al. Neuropilin-2 is a receptor for semaphorin IV: Insight into the structural basis of receptor function and specificity. Neuron 1998; 21:1079-1092.
8. Chedotal A, J.A. DR, Ruiz M et al. Semaphorins III and IV repel hippocampal axons via two distinct receptors. Development 1998; 125:4313-4323.
9. Chen H, Chedotal A, He Z et al. Neuropilin-2, a novel member of the neuropilin family, is a high affinity receptor for the semaphorins Sema E and Sema IV but not Sema III. Neuron 1997; 19:547-559.
10. Feiner L, Koppel AM, Kobayashi H et al. Secreted chick semaphorins bind recombinant neuropilin with similar affinities but bind different subsets of neurons in situ. Neuron 1997; 19:539-545.
11. Xu X, Ng S, Wu ZL et al. Human semaphorin K1 is glycosylphosphatidylinositol-linked and defines a new subfamily of viral-related semaphorins. J Biol Chem 1998; 273:22428-22434.
12. Takahashi T, Nakamura F, Jin Z et al. Semaphorins A and E act as antagonists of neuropilin-1 and agonists of neuropilin-2 receptors. Nature Neurosci 1998; 1:487-493.
13. Kobayashi H, Koppel AM, Luo Y et al. A role for collapsin-1 in olfactory and cranial sensory axon guidance. J Neurosci 1997; 17:8339-8352.

14. Takahashi T, Nakamura F, Strittmatter SM. Neuronal and non-neuronal collapsin-1 binding sites in developing chick are distinct from other semaphorin binding sites. J Neurosci 1997; 17:9183-9193.
15. Comeau MR, Johnson R, DuBose RF et al. A poxvirus-encoded semaphorin induces cytokine production from monocytes and binds to a novel cellular semaphorin receptor, VESPR. Immunity 1998; 8:473-482.
16. Winberg ML, Noordermeer JN, Tamagnone L et al. Plexin A is a neuronal semaphorin receptor that controls axon guidance. Cell 1998; 95:903-916.
17. Takahashi T, Fournier A, Nakamura F et al. Plexin-neuropilin-1 complexes form functional semaphorin-3A receptors. Cell 1999; 99:59-69.
18. Tamagnone L, Artigiani S, Chen H et al. Plexins are a large family of receptors for transmembrane, secreted, and GPI-anchored semaphorins in vertebrates. Cell 1999; 99:71-80.
19. Rohm B, Ottemeyer A, Lohrum M et al. Plexin/neuropilin complexes mediate repulsion by the axonal guidance signal semaphorin 3A. Mech Dev 2000; 93:95-104.
20. Maestrini E, Tamagnone L, Longati P et al. A family of transmembrane proteins with homology to the MET-hepatocyte growth factor receptor. Proc Nat Acad Sci USA 1996; 93:674-678.
21. Tamagnone L, Comoglio PM. Signalling by semaphorin receptors: cell guidance and beyond. Trends Cell Biol 2000; 10:377-383.
22. Bork P, Doerks T, Springer TA et al. Domains in plexins: links to integrins and transcription factors. Trends Biochem Sci 1999; 24:261-263.
23. Murakami Y, Suto F, Shimizu M et al. Differential expression of plexin-A subfamily members in the mouse nervous system. Dev Dyn 2001; 220:246-258.
24. Cheng HJ, Bagri A, Yaron A et al. Plexin-A3 mediates semaphorin signaling and regulates the development of hippocampal axonal projections. Neuron 2001; 32:249-263.
25. Bagnard D, Lohrum M, Uziel D et al. Semaphorins act as attractive and repulsive guidance signals during the development of cortical projections. Development 1998; 125:5043-5053.
26. Polleux F, Morrow T, Ghosh A. Semaphorin 3A is a chemoattractant for cortical apical dendrites. Nature 2000; 404:567-573.
27. Takahashi T, Strittmatter SM. PlexinA1 autoinhibition by the plexin sema domain. Neuron 2001; 29:429-439.
28. Rohm B, Rahim B, Kleiber B et al. The semaphorin 3A receptor may directly regulate the activity of small GTPases. FEBS Lett 2000; 486:68-72.
29. Vikis HG, Li W, He Z et al. The semaphorin receptor plexin-B1 specifically interacts with active rac in a ligand-dependent manner. Proc Natl Acad Sci USA 2000; 97:12457-12462.
30. Driessens MH, Hu H, Nobes CD et al. Plexin-B semaphorin receptors interact directly with active Rac and regulate the actin cytoskeleton by activating Rho. Curr Biol 2001; 11:339-344.
31. Hu H, Marton TF, Goodman CS. Plexin B mediates axon guidance in *Drosophila* by simultaneously inhibiting active rac and enhancing rhoA signaling. Neuron 2001; 32:39-51.
32. Zanata SM, Hovatta I, Rohm B et al. Antagonistic effects of Rnd1 and RhoD GTPases regulate receptor activity in Semaphorin 3A induced cytoskeletal collapse. J Neurosci 2002; in press.
33. Hall A. Rho GTPases and the actin cytoskeleton. Science 1998; 279:509-514.
34. Kozma R, Sarner S, Ahmed S et al. Rho family GTPases and neuronal growth cone remodelling: relationship between increased complexity induced by Cdc42Hs, Rac1, and acetylcholine and collapse induced by RhoA and lysophosphatidic acid. Mol Cell Biol 1997; 17:1201-1211.
35. Sander EE, ten Klooster JP, van Delft S et al. Rac downregulates Rho activity: reciprocal balance between both GTPases determines cellular morphology and migratory behavior. J Cell Biol 1999; 147:1009-1022.
36. Luo L. Rho GTPases in neuronal morphogenesis. Nat Rev Neurosci 2000; 1:173-180.
37. Wahl S, Barth H, Ciossek T et al. Ephrin-A5 induces collapse of growth cones by activating Rho and Rho kinase. J Cell Biol 2000; 149:263-270.

38. Shamah SM, Lin MZ, Goldberg JL et al. EphA receptors regulate growth cone dynamics through the novel guanine nucleotide exchange factor ephexin. Cell 2001; 105:233-244.
39. Jin Z, Strittmatter SM. Rac1 mediates collapsin-1-induced growth cone collapse. J Neurosci 1997; 15:6256-6563.
40. Kuhn TB, Brown MD, Wilcox CL et al. Myelin and collapsin-1 induce motor neuron growth cone collapse through different pathways: inhibition of collapse by opposing mutants of rac1. J Neurosci 1999; 19:1965-1975.
41. Västrik I, Eickholt BJ, Walsh FS et al. Sema3A-induced growth-cone collapse is mediated by Rac1 amino acids 17-32. Curr Biol 1999; 9:991-998.
42. Aizawa H, Wakatsuki S, Ishii A et al. Phosphorylation of cofilin by LIM-kinase is necessary for semaphorin 3A-induced growth cone collapse. Nat Neurosci 2001; 4:367-373.
43. Bagnard D, Vaillant C, Khuth ST et al. Semaphorin 3A-vascular endothelial growth factor-165 balance mediates migration and apoptosis of neural progenitor cells by the recruitment of shared receptor. J Neurosci 2001; 21:3332-33341.
44. Castellani V, Chedotal A, Schachner M et al. Analysis of the L1-deficient mouse phenotype reveals cross-talk between Sema3A and L1 signaling pathways in axonalguidance. Neuron 2000; 27:237-249.
45. Soker S, Takashima S, Miao HQ et al. Neuropilin-1 is expressed by endothelial and tumor cells as an isoform-specific receptor for vascular endothelial growth factor. Cell 1998; 92:735-745.

THE INTERACTION OF NEUROPILIN-1 AND NEUROPILIN-2 WITH TYROSINE-KINASE RECEPTORS FOR VEGF

Gera Neufeld*, Ofra Kessler and Yael Herzog

SUMMARY

The Neuropilin-1 (NRP1) and Neuropilin-2 (NRP2) receptors were initially described as receptors for axon guidance factors belonging to the class-3 Semaphorin sub-family. Subsequently, it was found the Neuropilins also function as receptors for some forms of vascular endothelial growth factor (VEGF). $VEGF_{165}$ binds to both NRP1 and to NRP2 but $VEGF_{121}$ does not bind to either of these receptors. $VEGF_{145}$ on the other hand, binds to NRP2 but not to NRP1. Additional VEGF family members such as the heparin binding form of placenta growth factor (PlGF-2) and VEGF-B bind to NRP1, and it was also shown that both PlGF-2 and VEGF-C bind to NRP2.

The intracellular domains of the Neuropilins are short, and do not suffice for independent transduction of biological signals subsequent to Semaphorin or VEGF binding. It was shown that both Neuropilins can form complexes with receptors belonging to the Plexin family, and that such Plexin/Neuropilin complexes are able to transduce signals following the binding of class-3 Semaphorins to Neuropilins. The $VEGF_{165}$ induced proliferation and migration of cells that express the VEGF tyrosine-kinase receptor VEGFR2 is enhanced in the presence of NRP1, suggesting that Neuropilins may also form complexes with VEGF tyrosine-kinase receptors such as VEGFR2. However, it is not yet clear whether VEGFR2 and NRP1 form complexes and contrasting results have been reported with regard to this issue. In contrast, it was recently reported by two laboratories that Neuropilins can form

*Department of Biology, Technion, Israel Institute of Technology, Haifa, 32000, Israel.

complexes with the second tyrosine-kinase receptor of VEGF, VEGFR1. However, the biological function of these complexes is still unclear.

INTRODUCTION

The A5 neuronal cell surface antigen was initially identified in xenopus embryos[1] and was subsequently renamed Neuropilin.[2] Neuropilin functions as a receptor for Semaphorin-3A (Sema-3A) which is one of the six axon repellent factors belonging to the class-III Semaphorin sub-family (reviewed in refs. 3,4). The class-3 Semaphorins induce the collapse of neuronal growth cones which is why they were initially named collapsins.[5] It was simultaneously found that yet another Neuropilin like gene was present in the human genome. Neuropilin was therefore renamed Neuropilin-1 (NRP1) and the related gene was named Neuropilin-2 (NRP2).[6,7] NRP2 was found to behave as a receptor for Semaphorin-3F (Sema-3F) which induces repulsion of NRP2 expressing neuronal growth cones, and for Semaphorin-3B (Sema-3B) and Semaphorin-3C (Sema-3C). NRP1 and NRP2 form homo and hetero-complexes[8] and the formation of such complexes is thought to be required for the transduction of Sema-3C signals.[9]

VEGF (also known as VEGF-A) is a major angiogenic factor that plays an essential role in embryonic vasculogenesis and angiogenesis.[10] At least five forms of VEGF are produced as a result of alternative splicing, and these forms differ with regard to the expression of exons 6 and 7 of the VEGF gene. Exons 6 and 7 encode independent heparin binding domains that are incorporated into longer VEGF forms. The shortest VEGF form, $VEGF_{121}$, lacks exons 6 and 7 altogether and does not bind to heparin. $VEGF_{165}$ includes the peptide encoded by exon 7, $VEGF_{145}$ includes the exon encoded by exon 6 and $VEGF_{189}$ includes both exons.[11] $VEGF_{121}$, $VEGF_{145}$ and $VEGF_{165}$ are secreted, and are active in cell proliferation assays and in angiogenesis assays.[10,12,13] In contrast, $VEGF_{189}$ displays a much higher affinity towards heparin and heparan-sulfate proteoglycans and is retained on cell surfaces.[13] However, it is interesting to note that $VEGF_{121}$ alone cannot compensate for the lack of other VEGF splice forms during embryonic development.[14]

All the VEGF splice forms bind to the VEGFR1 and to the VEGFR2 tyrosine-kinase receptors.[10] However, it was observed that human umbilical vein derived endothelial cells express VEGF receptors that were unable to bind $VEGF_{121}$ but bound $VEGF_{165}$. These cells express in addition VEGFR2 receptors but almost no VEGFR1 receptors.[15] Similar splice form specific receptors were subsequently found in several breast and prostate cancer derived cell lines which do not express VEGFR1 or VEGFR2.[16] Such cells were used as a source for the purification of these receptors, which were found to be the products of the NRP1 gene. It was observed that $VEGF_{165}$ did not have any effect upon cells that expressed NRP1 but lacked VEGFR1 or VEGFR2 even though $VEGF_{165}$ bound efficiently to NRP1 in such cells. In contrast, the $VEGF_{165}$ induced migration of cells that co-expressed VEGFR2 and Neuropilin-1 was enhanced as compared to the $VEGF_{165}$ induced migration of cells expressing VEGFR2 but no NRP1.[17] This was accompanied by an increase in the efficiency of

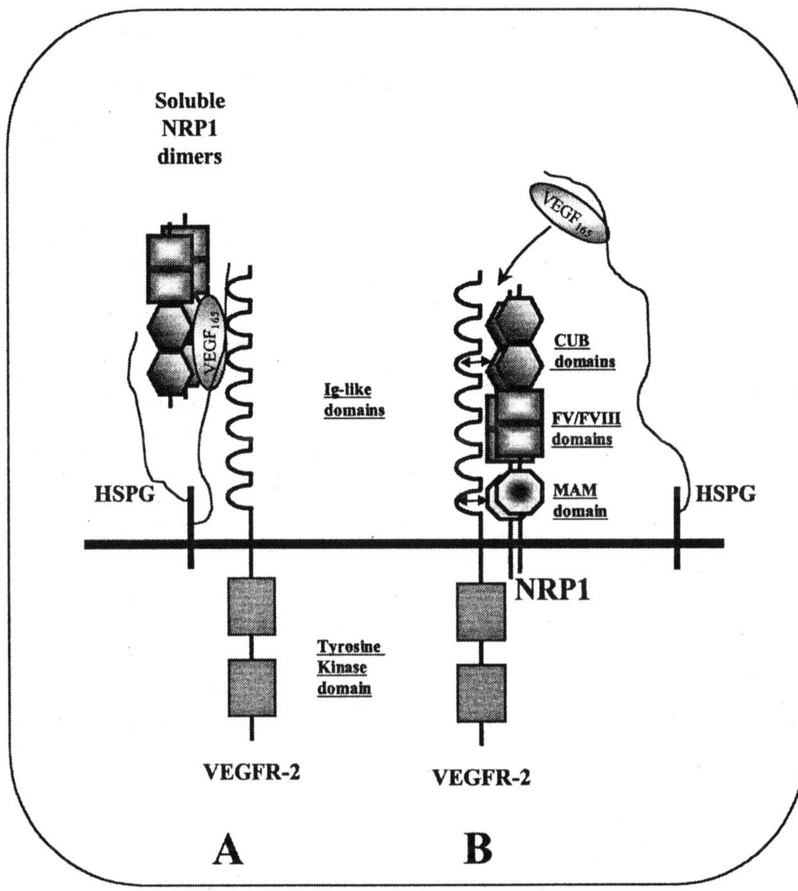

Figure 1. Alternative mechanisms by which NRP1 enhances $VEGF_{165}$ induced signal transduction via VEGFR2:
A. This mechanism postulates that NRP1 dimers which can be either soluble, anchored on the same cell, or anchored on adjacent cells bind $VEGF_{165}$ and subsequently present $VEGF_{165}$ to VEGFR2 receptors. Native soluble NPR1 lacks the MAM domain and is drawn as such.[33,34] The binding of $VEGF_{165}$ to NRP1 is enhanced by heparin and possibly by HSPG's. The binding of $VEGF_{165}$ to NRP1 dimers enhances the binding of $VEGF_{165}$ to VEGFR2, leading to an increase in the biological response to $VEGF_{165}$.[17] In this model no direct complexes are formed between VEGFR2 and NPR1.[22,25,34] Complex formation is indicated by double headed arrows (↔). The enhancing form of NRP1 is depicted as a homodimer based upon evidence which indicates that NRP1 dimers enhance while monomeric NRP1 inhibits $VEGF_{165}$ induced signaling via VEGFR2.[25] Motif names such as immuno-globulin like loops (Ig) are underlined.
B. In this model NPR1 dimers or monomers form complexes with VEGFR2 in the absence of $VEGF_{165}$. $VEGF_{165}$ binds to HSPG's or to heparin, which presents it to the pre-formed VEGFR2/NPR1 complex. The affinity of $VEGF_{165}$ to VEGFR2 is not affected but complex formation allows enhanced $VEGF_{165}$-induced signal transduction by VEGFR2 as compared to signal transduction by VEGFR2 in the absence of NRP1.[23]

$VEGF_{165}$ binding to VEGFR2 although why such an increase was observed was not clear. It was therefore concluded that NRP1 receptors cannot transduce VEGF signals on their own and that they probably function as accessory receptors that somehow enhance $VEGF_{165}$ induced signaling by VEGFR2.[17] It should be noted that these experiments were performed in the presence of heparin, a glycosaminoglycan which strongly enhances the interactions of $VEGF_{165}$ and of PlGF-2 with NRP1 and with NRP2 even in cells expressing endogenous heparan sulfate proteoglycans on their cell surfaces.[15,17-19]

The role of heparin in the interaction between $VEGF_{165}$ and Neuropilins is not very well understood. There is some evidence indicating that $VEGF_{165}$ binds to NRP1 via its heparin binding domain because peptides containing the heparin binding domain encoded by exon-7 of $VEGF_{165}$ inhibit the binding of $VEGF_{165}$ to NRP1.[20] These early experiments did not provide an explanation regarding the mechanism by which NRP1 affects signal transduction by VEGFR2. This mechanism was the subject of subsequent experiments as detailed in the next section.

THE MECHANISM BY WHICH NRP1 AFFECTS VEGF INDUCED SIGNALING BY THE VEGFR2 RECEPTOR

Two alternative mechanisms that explain how NRP1 enhances $VEGF_{165}$ induced signal transduction by VEGFR2 have been proposed. The first postulates that NRP1 binds and concentrates $VEGF_{165}$ on the cell surface, and presents it to VEGFR2 receptors. This mechanism would therefore result in an observable decrease of the dissociation constant between $VEGF_{165}$ and VEGFR2 as a result of the presence of NRP1. The second mechanism that was proposed assumes that NRP1 and VEGFR2 form complexes, and that the presence of NRP1 in these complexes enhances $VEGF_{165}$ induced signal transduction by VEGFR2. This mechanism does not assume necessarily that binding affinity of $VEGF_{165}$ to VEGFR2 increases, and postulates that signal transduction by VEGFR2/NRP1 complexes is more efficient than signal transduction by VEGFR2 alone.

These hypotheses were recently tested in two different studies. In the first study the interaction of $VEGF_{165}$ with the extracellular domains of NRP1 and VEGFR2 was examined using plasmon resonance.[21] In this study the dissociation constant of $VEGF_{165}$ binding to VEGFR2 was decreased by 3-5 fold in the presence of NRP1. The decrease did not seem to depend upon the formation of VEGFR2/NRP1 complexes and an interaction between the soluble extracellular domains of NRP1 and VEGFR2 could not be detected.[22] Thus, the results of this study appeared to support the first mechanism.

This last conclusion was challenged in another study which presented evidence for the existence of complexes containing VEGFR2 and NRP1. Antibodies directed against NRP1 were able to co-immunoprecipitate VEGFR2 from COS cells expressing both recombinant receptors but not from COS cells expressing VEGFR2 alone. Similar results were obtained when VEGFR2 specific antibodies were used.

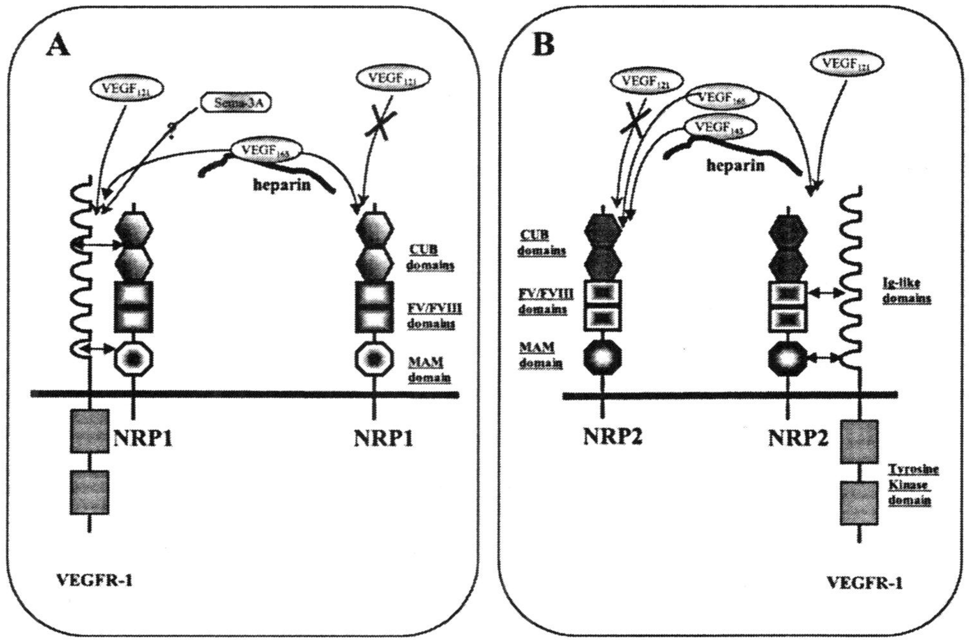

Figure 2. The interaction of NRP1 and NRP2 with VEGFR1:
A. The binding of $VEGF_{165}$ to NRP1 is strongly enhanced by heparin. Heparin binds to $VEGF_{165}$ and may also be able to interact with VEGFR-1. NRP1 probably forms a complex with VEGFR1 prior to the binding of $VEGF_{165}$.[22] Complex formation enables the binding of $VEGF_{121}$ to NRP1 while in the absence of VEGFR1, $VEGF_{121}$ fails to bind to NRP1.[31] Sema-3A may perhaps be able to bind to VEGFR1/ NRP1 complexes and to induce VEGFR-1 dependent signal transduction.[32] Complex formation is indicated by a double headed arrow (↔). Motif names such as immuno-globulin like loops (Ig) and tyrosine-kinase domains are underlined. Question mark symbolizes speculations for which there is no direct experimental proof.
B. NRP2 differs from NRP1 by its ability to bind $VEGF_{145}$ [19] but it too is unable to bind $VEGF_{121}$ on its own. NRP2 forms complexes with VEGFR-1 and the formation of these complexes enables the binding of $VEGF_{121}$ to NRP2.

Furthermore, the formation of these complexes appeared to be independent of the presence of $VEGF_{165}$. $VEGF_{165}$ seemed to bind to its binding sites on each of the receptors that form the complex since antibodies directed against VEGFR2 co-immunoprecipitated NRP1/^{125}I-$VEGF_{165}$ complexes along with VEGFR2/^{125}I-$VEGF_{165}$ complexes following the binding and cross-linking of ^{125}I-$VEGF_{165}$ to cells expressing both receptor types.[23] Contrary to the previous study in which it was observed that the affinity of $VEGF_{165}$ to VEGFR2 increased in the presence of NRP1[22] it was concluded that the affinity of $VEGF_{165}$ towards VEGFR2 does not change in the presence of NRP1.[23]

The differences between the results of these two studies are hard to reconcile. However, they were done using very divergent methods and reagents and may not be necessarily contradictory. One major difference is that in the first study soluble extracellular domains of the VEGF receptors were used[22] while in the second study the behavior of full length membrane bound receptors was examined.[23] It is possible that the intracellular domains of the receptors play a role in complex formation, and that this difference may account for the divergent results. The intracellular part of NRP1 may not be as inert with regard to protein-protein interactions. It was shown that NRP1 interacts with the NRP1 Interacting Protein (NIP) via a PDZ domain recognition sequence located at the C terminal of NRP1. NIP could function positively to link NRP1 with signaling molecules or the cytoskeleton thus affecting receptor clustering, it could also act as an inhibitory protein to mask critical regions of NRP1 that interact with other signaling molecules or receptors.[24]

An additional factor that may affect the interaction of NRP1 with VEGFR2 is the formation of NRP1 homo dimers or hetero dimers with NRP2. It was recently shown that soluble monomers of NRP1 extracellular domains inhibit $VEGF_{165}$ signaling mediated by VEGFR2 while NRP1 homodimers potentiate it.[25] These experiments, as well as experiments by additional researchers, indicate that NRP1 can affect VEGFR2 signaling even in trans,[26] and that the presence of NRP1 monomers or dimers may produce opposite effects upon $VEGF_{165}$ induced signal transduction via VEGFR2. The soluble NRP1 extracellular domains used in the first study were probably mainly monomeric[22] while the full length NRP1 receptors used in the second study probably formed homodimers,[23] and this difference may account for the contrasting results.

THE INTERACTION OF NEUROPILINS WITH VEGFR1

The biological role of the tyrosine-kinase receptor VEGFR1 (flt-1) has been an enigma since its characterization as a VEGF receptor.[27,28] Full length VEGFR1 has an essential role in the formation of the cardiovascular system as demonstrated by gene targeting experiments.[29] Surprisingly, it was found that a truncated VEGFR1 lacking the tyrosine-kinase domain is able to support the development of the cardiovascular system like the full length receptor.[30] As a result it was concluded that VEGFR1 probably fulfills an inhibitory role by sequestering VEGF, thus limiting VEGF activity. However, it is also possible that the effects of VEGFR1 upon the development of the cardiovascular system are mediated by complex formation between the extracellular domain of VEGFR1 and other cell surface receptors which may then transduce a signal.

We have noticed several years ago that human umbilical vein derived endothelial cells contain VEGF receptors that cannot bind the VEGF splice form $VEGF_{121}$.[15] These receptors were subsequently identified as the products of the Neuropilins.[17,19] However, human umbilical vein endothelial cells do not express, or express very little VEGFR1. We were therefore surprised to find in binding/cross-linking experiments that $VEGF_{121}$ was able to bind to both NRP1 and NRP2 in cells that co-express VEGFR1, suggesting that an interaction between VEGFR1 and the Neuropilins creates conditions that enable the binding of $VEGF_{121}$ to Neuropilins.[31] Co-immunoprecipitation experiments have shown that antibodies directed against VEGFR1 precipitated a cross-linked ^{125}I-VEGF/NRP2 complex and vice-versa. Thus we concluded that Neuropilins can form complexes with VEGFR1. The interaction may not be very stable since we have failed to detect immuno-percipitated complexes in experiments that were performed without prior cross-linking of ^{125}I-VEGF to the receptors, using just antibodies to detect precipitated receptors (although this failure may just represent a sensitivity problem) and we could not determine whether complex formation depended upon VEGF binding.[31]

Complex formation between the extracellular domains of VEGFR1 and NRP1 was also seen in experiments that employed plasmon resonance to detect complex formation.[22] In this study it was shown that a truncated extracellular domain of NRP1 (amino-acids 1-600) and the extracellular domain of VEGFR1 interact in the absence of VEGF, and the interaction is inhibited by heparin. The interaction depended upon the presence of a heparin binding domain in the VEGFR1 extracellular domain. This result suggests that high concentrations of VEGFR1 may sequester NRP1, especially under conditions in which the concentration of VEGF is limiting and in the absence of heparin-like molecules, leading to the impairment of efficient signal transduction by VEGFR2 by inhibiting the enhancing effect of NRP1 upon VEGFR2 signal transduction.[22]

The biological significance of complex formation between the Neuropilins and VEGFR1 remains unclear. However, it was recently reported that repulsion of migrating DEV neuronal progenitor cells by Sema-3A required, in addition to the presence of NRP1, the simultaneous presence of VEGFR1 receptors. Interestingly, both $VEGF_{165}$ and $VEGF_{121}$ inhibited the repulsive activity of Sema-3A. These results indicate that Sema-3A may be able to interact with VEGFR-1 directly, or alternatively, that Sema-3A can interact with a VEGFR1/NRP1 complex. Either possibility accounts for the observed inhibition of the Sema-3A induced effect by $VEGF_{121}$.[31,32] Although the formation of complexes between VEGFR1 and NRP1 was not examined in this study, it nevertheless suggests strongly that the formation of such complexes may also play a role in Semaphorin induced signal transduction in certain cell types, and that the function of VEGFR1 may not be restricted to the cardiovascular system.

CONCLUSIONS

Several studies have provided evidence for the formation of complexes between the Neuropilins and tyrosine-kinase receptors of VEGF. These are surprising observations since these tyrosine-kinase receptors can evidently bind VEGF and transduce VEGF signals without assistance. In contrast, Plexins cannot bind class-3 Semaphorins, and Neuropilins are required as the ligand binding part in the holoreceptors which they form. What than is the benefit derived from the interaction of an autonomous tyrosine-kinase receptor with Neuropilins? The most obvious explanation is fine-tuning. By interacting with Neuropilins the activities of the tyrosine-kinase receptors may be modulated to suit specific conditions. However, it is also possible that Neuropilins may serve as the nuclei for the formation of signaling complexes containing more than two distinct components. It is possible for example, that such complexes may contain a VEGF tyrosine-kinase receptor, a Neuropilin and a Plexin. In such a way it may perhaps be possible to induce Plexin mediated signaling by VEGF, and thereby enable cross-talk between seemingly unrelated signal transduction pathways. More experiments will be required to examine such possibilities.

REFERENCES

1. Takagi S, Hirata T, Agata K, Mochii M, Eguchi G, Fujisawa H. The A5 antigen, a candidate for the neuronal recognition molecule, has homologies to complement components and coagulation factors. Neuron 1991; 7(2):295-307.
2. Fujisawa H, Takagi S, Hirata T. Growth-associated expression of a membrane protein, neuropilin, in Xenopus optic nerve fibers. Dev Neurosci 1995; 17(5-6):343-349.
3. Raper JA. Semaphorins and their receptors in vertebrates and invertebrates. Curr Opin Neurobiol 2000; 10(1):88-94.
4. Goodman CS, Kolodkin AL, Luo Y, Pueschel AW, Raper JA. Unified nomenclature for the semaphorins collapsins. Cell 1999; 97(5):551-552.
5. Luo Y, Raible D, Raper JA. Collapsin: a protein in brain that induces the collapse and paralysis of neuronal growth cones. Cell 1993; 75(2):217-227.
6. He Z, Tessier-Lavigne M. Neuropilin is a receptor for the axonal chemorepellent Semaphorin III. Cell 1997; 90(4):739-751.
7. Kolodkin AL, Levengood DV, Rowe EG, Tai YT, Giger RJ, Ginty DD. Neuropilin is a semaphorin III receptor. Cell 1997; 90(4):753-762.
8. Giger RJ, Urquhart ER, Gillespie SK, Levengood DV, Ginty DD, Kolodkin AL. Neuropilin-2 is a receptor for semaphorin IV: Insight into the structural basis of receptor function and specificity. Neuron 1998; 21(5):1079-1092.
9. Takahashi T, Nakamura F, Jin Z, Kalb RG, Strittmatter SM. Semaphorins A and E act as antagonists of neuropilin-1 and agonists of neuropilin-2 receptors. Nat Neurosci 1998; 1:487-493.
10. Neufeld G, Cohen T, Gengrinovitch S, Poltorak Z. Vascular endothelial growth factor (VEGF) and its receptors. FASEB J 1999; 13:9-22.
11. Robinson CJ, Stringer SE. The splice variants of vascular endothelial growth factor (VEGF) and their receptors. J Cell Sci 2001; 114(Pt 5):853-865.
12. Poltorak Z, Cohen T, Sivan R, Kandelis Y, Spira G, Vlodavsky I et al. $VEGF_{145}$: A secreted VEGF form that binds to extracellular matrix. J Biol Chem 1997; 272:7151-7158.

13. Park JE, Keller GA, Ferrara N. Vascular endothelial growth factor (VEGF) isoforms—Differential deposition into the subepithelial extracellular matrix and bioactivity of extracellular matrix-bound VEGF. Mol Biol Cell 1993; 4:1317-1326.
14. Carmeliet P, Ng YS, Nuyens D, Theilmeier G, Brusselmans K, Cornelissen I et al. Impaired myocardial angiogenesis and ischemic cardiomyopathy in mice lacking the vascular endothelial growth factor isoforms $VEGF_{164}$ and $VEGF_{188}$. Nature Med 1999; 5(5):495-502.
15. Gitay-Goren H, Cohen T, Tessler S, Soker S, Gengrinovitch S, Rockwell P et al. Selective binding of VEGF121 to one of the three VEGF receptors of vascular endothelial cells. J Biol Chem 1996; 271:5519-5523.
16. Soker S, Fidder H, Neufeld G, Klagsbrun M. Characterization of novel VEGF binding proteins associated with tumor cells that bind VEGF165 but not VEGF121. J Biol Chem 1996; 271:5761-5767.
17. Soker S, Takashima S, Miao HQ, Neufeld G, Klagsbrun M. Neuropilin-1 is expressed by endothelial and tumor cells as an isoform specific receptor for vascular endothelial growth factor. Cell 1998; 92:735-745.
18. Migdal M, Huppertz B, Tessler S, Comforti A, Shibuya M, Reich R et al. Neuropilin-1 is a placenta growth factor-2 receptor. J Biol Chem 1998; 273(35):22272-22278.
19. Gluzman-Poltorak Z, Cohen T, Herzog Y, Neufeld G. Neuropilin-2 and Neuropilin-1 are receptors for 165-amino acid long form of vascular endothelial growth factor (VEGF) and of placenta growth factor-2, but only neuropilin-2 functions as a receptor for the 145 amino acid form of VEGF. J Biol Chem 2000; 275:18040-18045.
20. Soker S, Gollamudi-Payne S, Fidder H, Charmahelli H, Klagsbrun M. Inhibition of vascular endothelial growth factor (VEGF) induced endothelial cell proliferation by a peptide corresponding to the exon-7 encoded domain of VEGF165. J Biol Chem 1997; 272(50):31582-31588.
21. McDonnell JM. Surface plasmon resonance: towards an understanding of the mechanisms of biological molecular recognition. Curr Opin Chem Biol 2001; 5(5):572-577.
22. Fuh G, Garcia KC, De Vos AM. The interaction of Neuropilin-1 with vascular endothelial growth factor and its receptor Flt-1. J Biol Chem 2000; 275(35):26690-26695.
23. Whitaker GB, Limberg BJ, Rosenbaum JS. Vascular endothelial growth factor receptor-2 and neuropilin-1 form a receptor complex that is responsible for the differential signaling potency of $VEGF_{165}$ and $VEGF_{121}$. J Biol Chem 2001; 276(27):25520-25531.
24. Cai HB, Reed RR. Cloning and characterization of neuropilin-1-interacting protein: A PSD-95/Dlg/ZO-1 domain-containing protein that interacts with the cytoplasmic domain of neuropilin-1. J Neurosci 1999; 19(15):6519-6527.
25. Yamada Y, Takakura N, Yasue H, Ogawa H, Fujisawa H, Suda T. Exogenous clustered neuropilin 1 enhances vasculogenesis and angiogenesis. Blood 2001; 97(6):1671-1678.
26. Miao HQ, Lee P, Lin H, Soker S, Klagsbrun M. Neuropilin-1 expression by tumor cells promotes tumor angiogenesis and progression. FASEB J 2000; 14(15):2532-2539.
27. Devries C, Escobedo JA, Ueno H, Houck K, Ferrara N, Williams LT. The fms-like tyrosine kinase, a receptor for vascular endothelial growth factor. Science 1992; 255:989-991.
28. Shibuya M, Yamaguchi S, Yamane A, Ikeda T, Tojo A, Matsushime H et al. Nucleotide sequence and expression of a novel human receptor type tyrosine kinase gene (flt) closely related to the fms family. Oncogene 1990; 5:519-524.
29. Fong GH, Rossant J, Gertsenstein M, Breitman ML. Role of the Flt-1 receptor tyrosine kinase in regulating the assembly of vascular endothelium. Nature 1995; 376:66-70.
30. Hiratsuka S, Minowa O, Kuno J, Noda T, Shibuya M. Flt-1 lacking the tyrosine kinase domain is sufficient for normal development and angiogenesis in mice. Proc Natl Acad Sci USA 1998; 95(16):9349-9354.
31. Gluzman-Poltorak Z, Cohen T, Shibuya M, Neufeld G. Vascular endothelial growth factor receptor-1 and neuropilin-2 form complexes. J Biol Chem 2001; 276(22):18688-18694.

32. Bagnard D, Vaillant C, Khuth ST, Dufay N, Lohrum M, Puschel AW et al. Semaphorin 3A-vascular endothelial growth factor-165 balance mediates migration and apoptosis of neural progenitor cells by the recruitment of shared receptor. J Neurosci 2001; 21(10):3332-3341.
33. Gagnon ML, Bielenberg DR, Gechtman Z, Miao HQ, Takashima S, Soker S et al. Identification of a natural soluble neuropilin-1 that binds vascular endothelial growth factor: In vivo expression and antitumor activity. Proc Natl Acad Sci USA 2000; 97(6):2573-2578.
34. Miao HQ, Klagsbrun M. Neuropilin is a mediator of angiogenesis. Cancer Metastasis Rev 2000; 19(1-2):29-37.

THE FUNCTION OF NEUROPILIN / L1 COMPLEX

V. Castellani*

SUMMARY

L1, a cell adhesion molecule of the Ig superfamily (IgCAM) plays a critical role in the formation of neuronal networks. This is reflected by the variety of clinical signs associated with the X-linked recessive neurological disorder that is caused by mutations in the L1 gene. L1 regulates the formation of axon fascicles and promotes neurite outgrowth through interaction with a wide spectrum of binding partners including cell adhesion molecules and extra-cellular matrix components. Here we describe the emerging evidence that indicates, in addition to these well-established functions, that L1 participates in the signaling of a secreted guidance cue of the Semaphorin family, Sema3A. Three types of experimental evidence support L1 as a key component of the Sema3A receptor complex. First, L1-deficient axons do not respond to Sema3A-induced chemorepulsion. Second, L1 and NRP1, the neuropilin responsible for Sema3A binding, associate through their extracellular domains, forming a cell surface heterocomplex. Third, a soluble form of L1 modulates axonal responsiveness to Sema3A, by converting Sema3A chemorepulsion into attraction.

INTRODUCTION

It has become clear over the last few years that secreted semaphorins activate multimolecular receptor complexes that transduce a repulsive or attractive signal to the growth cone.[1] So far, the two main components of class III Semaphorin receptors that have been characterized are members of the Neuropilin (NRP) and Plexin families. Neuropilin 1 (NRP1) and Neuropilin 2 (NRP2) are responsible for ligand

*Laboratoire de Neurogenése et Morphogenése dans le Développement et chez l'Adulte, UMR 6156, Université de la Méditerranée, IBDM, Parc Scientifique de Luminy, 13288 Marseille cedex 9, France.

binding in the complex whereas Plexins transduce the semaphorin signal by coupling it to the internal cytoskeletal dynamic of the growth cone. The secreted semaphorins display particular features, for example Sema3F binds both NRPs however only one NRP is required to repel the growth cones. Furthermore in the receptor complex, several different plexins can induce a repulsive response for a single semaphorin.[1] In NRP1-expressing cells, Plex-A1, Plex-A2 and Plex-A3 (although only partially, see ref. 2) confer a cellular response for Sema3A.[1] Conversely, when associated with the appropriate NRP, a plexin can transduce more than one semaphorin signal (i.e., Plex-A1 transduces a signal for Sema3A and Sema3F, see Ref. 1). Since neither NRPs nor plexins appear to be selective for specific semaphorins, it remains unclear how/whether the response to individual members of this family is indeed specified at the level of the receptor complex. One possibility is that different combinations of plexins may form specific receptor complexes, alternatively additional components of the complex may themselves be specific for each semaphorin. Recent findings described in this Chapter favor the latter hypothesis as they demonstrate that L1, a cell adhesion molecule of the Ig superfamily, associates with NRP1 and is selectively required for axonal responses to one of the class III Semaphorins, Sema3A.

L1 AND NRP1 ASSOCIATE THROUGH THEIR EXTRACELLULAR DOMAINS

L1, NrCAM, CHL1 and Neurofascin form a sub-group of neural IgCAMs that share highly conserved cytoplasmic region. These regions contain protein binding sites for interaction with the actin cytoskeleton and signaling cascades such as the MAP Kinase and the FGF receptor pathways.[3,4] IgCAMs play an important role in a number of functions during the development of the nervous system these include cell migration, neurite outgrowth, axon guidance and fasciculation (For a review see Ref. 5). In addition, in adulthood they contribute to the functioning of neuronal networks by regulating structural changes associated to synaptic plasticity.[6] This wide spectrum of biological activities is a result of a very complex pattern of homophilic and heterophilic binding with IgCAMs and extracellular matrix molecules (Figure 1; ref. 5).

The formation of a molecular complex between L1 and NRP1 has been observed using a biochemical approach. L1 and NRP1 were co-expressed transiently in COS7 cells and the cell lysate was immunoprecipitated with anti-L1 or anti-NRP1 antibodies. L1 and NRP1 were both found to co-precipitate. In contrast, no complex formation of L1 with NRP2 was observed using the same technique. This finding indicated that when L1 and NRP1 are present on the same cell membrane, they are able to form a complex. However, it was not known whether this complex was formed under physiological conditions. Thus, immunoprecipitation experiments were conducted on brain extracts, prepared from mice at postnatal developmental stage. Western blot analysis revealed a co-precipitation of L1 and NRP1, demonstrating that the complex formation occurs in vivo. To determine whether L1 and NRP1 associate directly within the complex a soluble form of L1, composed of the extracellular domain of the protein fused to the Fc fragment of the human immunoglobulin (L1Fc), was incubated with NRP1-expressing COS7 cells. Immunodetection of L1Fc with anti-Fc antibodies showed that the chimera bound to NRP1 but not NRP2 expressing cells. These finding demonstrated a direct association between the extracellular domains of L1 and NRP1. What could be the function of L1/NRP1 complex formation in the developing brain? The following paragraphs describe several sets of experiments suggesting that the function of the L1/NRP1 complex is to regulate axonal responses to the chemorepellent Sema3A during the formation of a specific cortical tract that establishes connections between the cerebral cortex and the spinal cord.

L1/NRP1 COMPLEX FORMATION REGULATES AXONAL RESPONSIVENESS TO A SECRETED SEMAPHORIN

In 1994, the significance of the role of L1 in the formation of the nervous system was highlighted by the identification of a direct causal relationship between an X-linked human neurological disease and the gene encoding L1. Mutations in the L1 gene cause a pathology referred to as X-linked hydrocephalus, or MASA syndrome (for Mental retardation, Aphasia, Shuffling gait, Adducted thumbs). Survivors exhibit morphological anomalies resulting from an abnormal development of some brain structures and neural fiber tracts.[7-9] Null-mutant mouse models have been generated to further understand the specific defects in neural development that are responsible for the human disease.[10,11] Strikingly L1-deficient mice exhibit many of the defects observed in the human brain, in particular aberrant development of the corticospinal tract (CST). The CST is a long-range projection primarily arising from all areas of the cerebral cortex that extends into the spinal cord. Extensive elimination of exuberant projections from inappropriate cortical areas over the first postnatal weeks leads to a restricted mature pattern that precisely connect the motor cortex to spinal cord motoneurons and interneurons, generating functions in the control of voluntary superior limb movements.[12,13] In vitro experiments conducted on the L1-deficient mice suggest that L1/NRP1 complex formation is involved in the guidance of CST axons at one of the most critical steps of its pathfinding, the pyramidal decussation.[14]

The developing CST leaves the cerebral cortex and travels ventrally through the medulla until it reaches the spinal cord at a stage corresponding to the first postnatal day in the mouse. At this level, axons turn to cross the midline and join the dorsal funiculus a process referred to as the pyramidal decussation (Figure 2A). Analysis of the L1 mutant phenotype with axonal tracers injected into the somatomotor cortex demonstrated that L1 is one of the guidance cues controlling the decisions of the growth cone at the decussation.[11] In the mutant, corticospinal axons either fail to cross the midline or they cross it but stay ventrally instead of

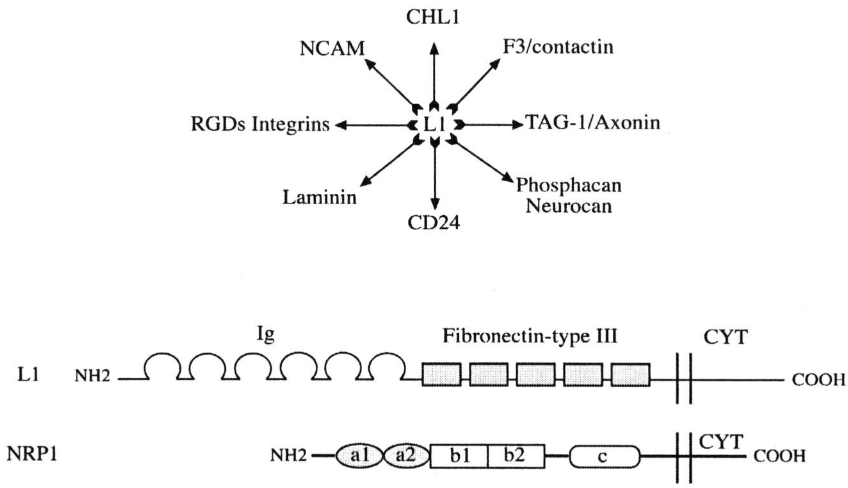

Figure 1. L1 displays a highly complex pattern of interactions with many different IgCAMs and also extracellular matrix molecules. Organization of L1 and NRP1 proteins: the extracellular domain of L1 is composed of 6 Ig domains and 5 Fibronectin type III domains. L1 cytoplasmic domain contains binding motifs for proteins associated with the actin cytoskeleton. NRP1 contains 2 CUB domains (a1 and a2), 2 factor V/VIII coagulation domains, a juxtamembrane MAM domain and a short cytoplasmic tail. Ig: immunoglobulin domain; CYT: cytoplasmic domain.

ascending towards the dorsal spinal cord (ref. 11; Figure 2A). Co-cultures of cortical slices and spinal cord explants (Figure 2B) were performed to investigate whether chemotropic mechanisms trigger the change of axon trajectory at the decussation.[14] This study revealed that cells residing in the ventral spinal cord secrete a repellent factor to cortical axons that belongs to the semaphorin family as the chemorepulsive axonal response was efficiently blocked by application of anti-NRP1 antibodies. Axons lacking L1 (extending from L1-deficient cortical slices) totally failed to respond to this chemorepellent, suggesting that in normal development, the L1/NRP1 complex may form to allow the growth cone to integrate the repulsive semaphorin message emanating from the spinal cord tissue. Members of the IgCAM superfamily were already known to be components of receptor complexes for other chemotropic signals, these include the receptors for the Netrin and Slit families DCC and Robo respectively.[15,16] However, a role of L1 in semaphorin signaling was unexpected for two main reasons. First, in contrast to DCC and Robo for which no obvious roles other than the signaling of guidance cues have been so far identified, L1 was already found to regulate a variety of biological functions. Second, as mentioned in the above paragraphs, L1 was not known to regulate any mechanisms other than cell-contact.

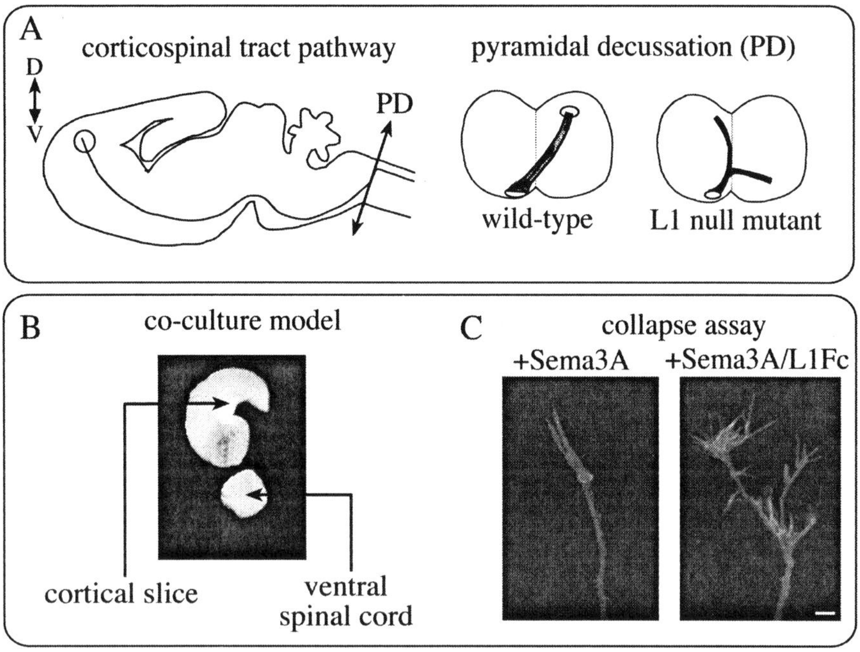

Figure 2. A) Schematic representation of a brain sagittal section showing the ventral pathway of the corticospinal tract and the pyramidal decussation, when axons enter the cervical spinal cord. Schemes of coronal sections at the pyramidal decussation to illustrate the guidance defects observed by Cohen et colleagues (Ref. 11) in the L1 null mutant. Instead of growing from the ventral to the dorsal column of the spinal cord and crossing the midline as in the wild-type animal, axons stay ipsilaterally and ventrally in the L1 null mutant. B-C) Microphotographs illustrating the co-culture and the collapse assays developed for investigating the chemotropic guidance at the pyramidal decussation. (B) In the co-culture model, a cortical slice is cultured with a ventral spinal cord explant, prepared from the cervical spinal cord, at the junction with the caudal medulla. The influence of the spinal cord explant on the trajectory of axons extending from the cortical slice is analyzed. (C) In the collapse assay, cortical slices are cultured until axons extend out of the slices and are exposed to Sema3A, either alone or together with L1Fc. The panels show phalloïdin-FITC staining to visualize filopodial and lamellipodial structures of the growth cone. These actin cytoskeletal structures are disorganized when the growth cone contacts Sema3A. L1Fc applied in combination with Sema3A protects the growth cone structures from the Sema3A-induced collapse. D:Dorsal; V:Ventral.

L1/NRP1 COMPLEX SPECIFIES GROWTH CONE RESPONSES TO SEMA3A

The results from the co-culture experiments raised the question of whether L1/NPR1 association dictates the growth cone response to one particular semaphorin or whether it is required for several. The question was addressed using co-cultures of cortical slices and COS cell aggregates secreting individual Semaphorins. In this culture model, axonal responses of wild-type and L1-deficient axons were compared in order to identify semaphorin(s) that would only induce a response with wild-type but not L1-deficient axons. Among the Semaphorins tested, only one of them, Sema3A, fulfilled these criteria, all others repelled to the same extend wild-type and L1-deficient axons.

Why L1/NRP1 is specifically required for Sema3A signaling and not for other semaphorins that also bind this NRP is still an open question. Perhaps it is due to the fact that Sema3A binds exclusively to NRP1 and uses it to induce a guidance response. One of the most obvious questions raised by the observation that L1-deficient axons are unresponsive to Sema3A is whether L1/NRP1 complex forms a functional receptor on its own or whether L1 is an additional component of NRP1-Plexin receptor complex. To answer this question, it will be necessary not only to investigate the interactions between L1 and Plexins, but more importantly to block the plexin signal in cortical neurons and examine the Sema3A-induced response. Another question is the cell-type specificity of L1-NRP1 complex. Does it only occur in cortical neurons or is it a general pre-requisite for Sema3A to elicit a response? A first answer was given by investigating the behavior of a second cell population known to be Sema3A-responsive, the DRG (Dorsal root ganglia) neurons. In the co-culture assay, wild-type but not L1-deficient DRG axons were repelled by Sema3A, indicating that L1-NRP1 complex formation is not restricted to cortical neurons.[14] Consequently, it is possible that DRG projections are altered by the genetic invalidation of L1.

An important issue to address is whether IgCAMs are required for other secreted semaphorins to generate a growth cone response. Intriguingly, a recent work reported that the function of a chemorepellent for dorsal root ganglia axons secreted from the notochord was diminished by application of antibodies directed against the GPI anchor IgCAM Axonin.[17] Moreover, enzymatic removal of GPI-anchor proteins from the axons also reduced the chemorepulsive response. Although the nature of the secreted factor remains unknown, these data strongly argue for other cross-talks between IgCAMs and chemorepulsive cues to control axon guidance responses.

SOLUBLE L1 MODULATES AXONAL RESPONSIVENESS TO SEMA3A

To further explore the functions of the L1/NRP1 complex formation, the influence of a soluble form of L1 on axonal responses to Sema3A was examined in collapse assays. These assays required culturing cortical slices until axons emerge from the tissue and then applying either Sema3A alone or in combination with the soluble L1Fc chimera for one hour. After fixation, the morphology of the growth cones was examined. Strikingly, L1Fc was able to prevent cortical growth cones from collapsing in response to Sema3A (Figure 2C) but as expected from the co-culture experiments, it did not affect Sema3B-induced collapse. Further evidence to support this property of the soluble L1Fc chimera to modify the sensitivity of the growth cones to Sema3A was obtained using co-cultures of cortical slices and ventral spinal cord. In the absence of treatment, L1Fc not only abrogated the ventral spinal cord-induced chemorepulsion observed but in fact switched repulsion to attraction. As these explants may also produce additional guidance cues, a possible interpretation was that rather than switching Sema3A to attraction, soluble L1 inhibits Sema3A-induced chemorepulsion, thereby modifying the balance of effects in favor of chemoattraction. However, a similar switching of axonal response was obtained in a simplest model whereby cortical slices were cultured with COS cell aggregates secreting only Sema3A. Thus, soluble L1 converts the chemorepulsive properties of Sema3A into attraction. Figure 3 summarizes axonal behaviors in response to Sema3A, depending on L1/NRP1 interactions on the growth cones.

Switching axonal responses to guidance cues was established earlier by Poo and colleagues and this process depends upon internal levels of cyclic nucleotides.[18] Consistently, soluble L1 activates the synthesis of NO and cGMP in order to exert its effects (unpublished observations). What is the biological relevance of these modulatory effects of soluble L1? Proteolytic sites for plasminogen and ADAM metalloprotease activities have been identified within the extracellular domain of the integral L1 although such enzymatic processing was not very clearly known to occur in vivo.[19,20] A recent study demonstrated the presence of cleaved/soluble L1 in the developing brain, peaking during the postnatal stages.[21] In the context of axon guidance, the release of L1 may on the one hand stop the growth cone response to Sema3A but on the other hand it may switch the repulsion of neighboring growth cones to attraction. An important issue will be to determine whether the modulation of axon responsiveness to Sema3A by soluble L1 is indeed required to guide developing neuronal projections in vivo.

OTHER PUTATIVE FUNCTIONS SERVED BY L1/NRP1 COMPLEX FORMATION

In the developing nervous system, axons navigate in bundles (except the earliest axons that pioneer the pathway). This selective fasciculation distinguishes axons destined to innervate common targets from those that will project on other tissues, a mechanism that largely contributes to pattern neuronal projections.[22] An example of this is the thalamocortical afferents which navigate in tight bundles until they reach the cerebral cortex where they segregate according to which thalamic nuclei they originate from and which cortical area they will project to.[23,24] Similarly, mixed pools of motor neuron axons initially form common bundles to exit the spinal cord and secondly defasciculate in the plexus so as to re-organize according to their muscle targets.[25] This is the type of event that might be regulated by L1/NRP1 complex formation, as both proteins mediate cell adhesion. The original function attributed to the NRP1 protein was indeed the regulation of axon fasciculation, as it was expressed along neuronal projections (see Chapter 1).[26,27] The analysis of the NRP1 null mutant showing defasciculation of PNS efferent projections further demonstrated the property of NRP1 to control axon bundling.[28] Defasciculation defects have also been proposed for explaining the CST hypoplasia observed at the level of

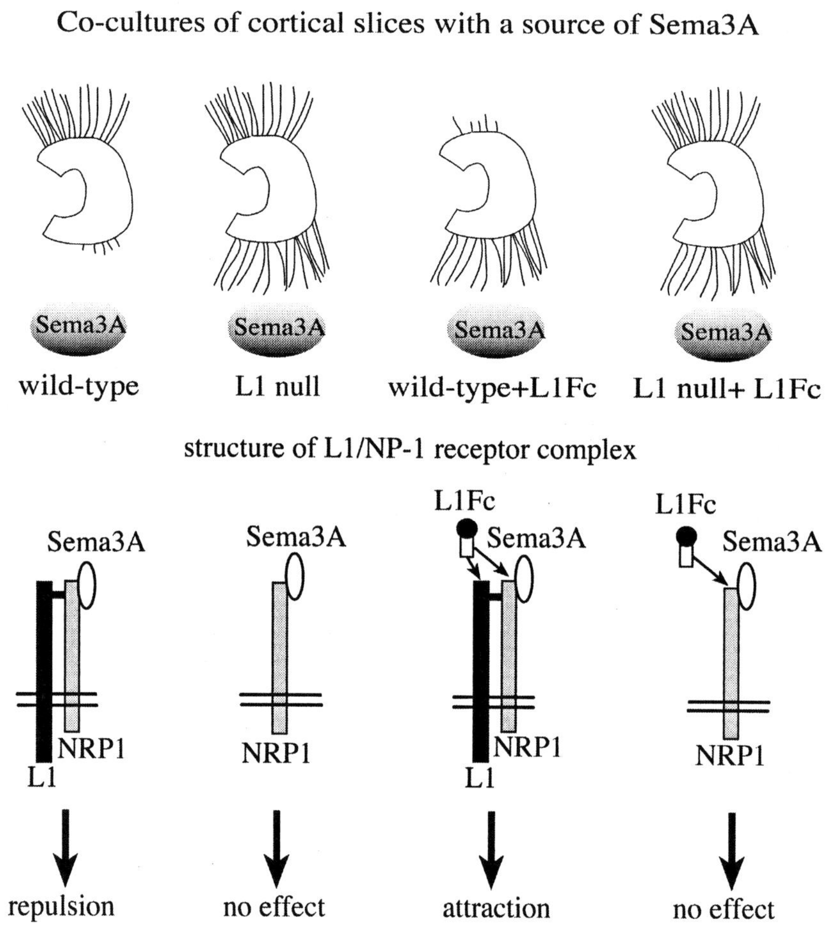

Figure 3. Schematic illustration of axon behaviors in the co-culture model correlated to L1/NRP1 interactions in the Sema3A receptor complex. Axons extending from wild-type cortical slices are repelled by Sema3A, and this response is switched to attraction by application of soluble L1 (L1Fc). In contrast, axons extending from L1-deficient slices neither are repelled nor are attracted by Sema3A, probably because the receptor complex is not functional due to the lack of L1.

the caudal medulla in the L1 null mutant.[10] Therefore L1/NRP1 *cis* association could control some of the interactions contracted by L1 and NRP1 with other axonal molecules in the bundles. Alternatively, L1/NRP1 *trans* interaction could mediate a selective recognition of subpopulations of axons that would express both proteins in a reciprocal pattern.

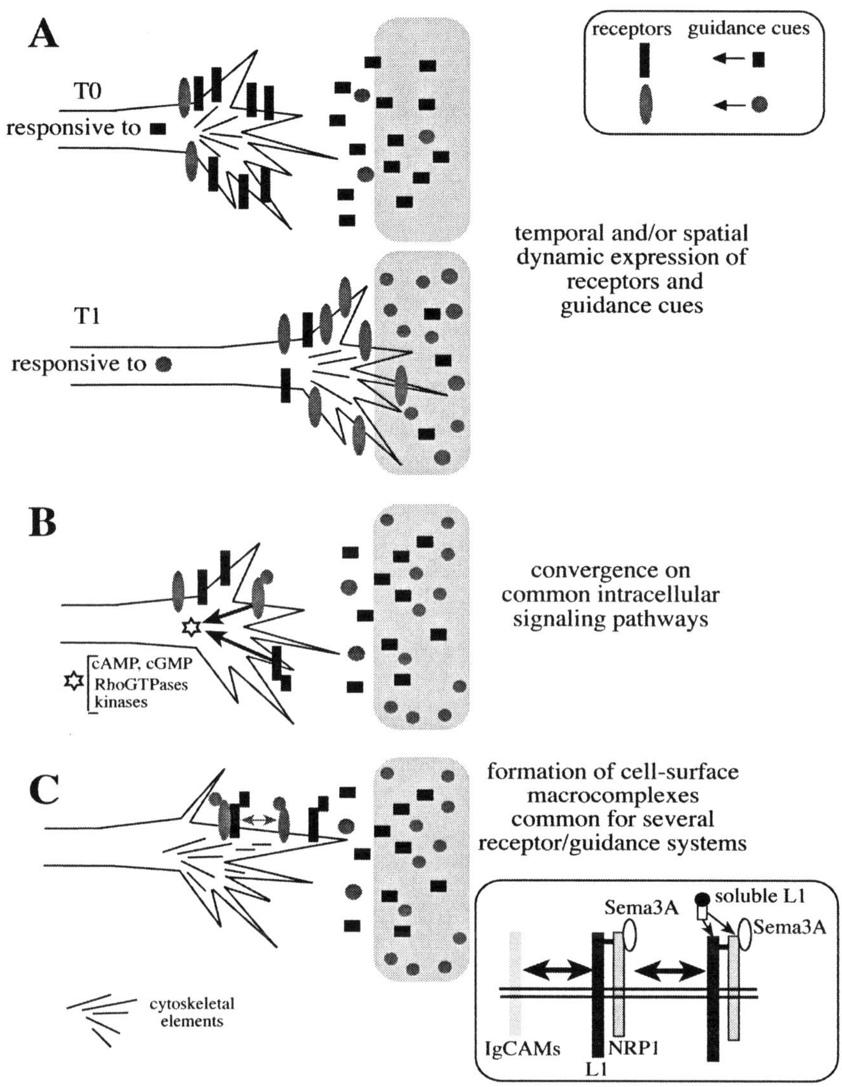

Figure 4. Mechanisms ensuring the coordination of growth cone responses. A) The dynamic expression of receptors and guidance cues allows the growth cone to become responsive to a guidance cue or conversely to be desensitized to it. B) Downstream effectors can be shared by different classes of guidance factors. Specific cascades initiated by distinct cues will therefore converge on common intracellular targets, and the growth cone response will finally depend on a net balance reflecting the different guidance components. C) A third possible mechanism is the association of different types of receptors for guidance cues. In this model, signaling pathways will differ depending on the composition or the conformation of the receptor units at the cell surface. The association of two types of receptors either inactivates or modifies the signaling triggered by one or both individual receptors. For example, L1/NRP1 association in the Sema3A receptor complex is required for the repulsive signal. When soluble L1 binds to L1 and NRP1, it modifies the receptor structure and therefore the intracellular cascade. Association of L1 with other IgCAMs expressed on growth cones is another potential way for modulating the Sema3A signaling.

PIVOTAL MOLECULES IN AXON GUIDANCE

In the developing nervous system, long-range projections might often undergo complex changes in their trajectory to reach their final target tissue. The crossing of the midline by sub-sets of axons connecting the two halves of the body is one of the best examples illustrating the complexity of the mechanisms regulating growth cone decisions and has been the subject of several recent reviews.[29,30] It is orchestrated by a combination of cues including IgCAMs, and members of the secreted Slit, Netrin and Semaphorin families. How do multiple positional cues generate a unique and coordinated growth cone response? It arises firstly through a highly complex series of up- and down-regulation of axonal receptors and guidance factors themselves. Second, the transduction pathways converge on downstream targets common to large classes of guidance cues in such a way that the growth cone decision will finally depend on a limited number of internal effectors (cyclic nucleotides and rhoGTPases for example, see references 31,32). Third, the formation of cell surface macrocomplexes that are common for sensing diverse classes of guidance cues is another mechanism for coordinating growth cone behaviors. An example of this can be seen in the *cis* association between the two "ligand-independent" receptors Robo1 (binding Slit-1) and DCC (binding Netrin-1) that induces a silencing of the Netrin signal.[33] *Cis* and *trans* interactions between L1 and NRP1 would fall into this type of regulatory mechanism allowing the coordination of cell-cell contact and secreted information through L1-dependent structural changes in the Sema3A receptor complex (See Figure 4). These modulations could be even more complex since L1 has other binding partners at the cell surface and these additional *cis* interactions could interfere with the Sema3A signal. Therefore, a central issue will be to understand how structural changes of these macrocomplexes shift the intracellular signaling pathways and subsequently the growth cone responses.

ACKNOWLEDGMENTS

I thank Elena De Angelis and Julien Falk for helpful comments on the manuscript.

REFERENCES

1. Tamagnone L, Comoglio PM. Signalling by semaphorin receptors: cell guidance and beyond.Trends Cell Biol 2000;10, 377-383.
2. Cheng HJ, Bagri A, Yaron A et al. Plexin-a3 mediates semaphorin signaling and regulates the development of hippocampal axonal projections. Neuron 2001 25;32(2):249-263.
3. Kamiguchi H, Lemmon V. IgCAMs: bidirectional signals underlying neurite growth.Curr Opin Cell Biol 2000;12(5):598-605.
4. Doherty P, Williams G, Williams EJ. CAMs and axonal growth: a critical evaluation of the role of calcium and the MAPK cascade. Mol Cell Neurosci 2000;16(4):283-295.
5. Brummendorf T, Rathjen FG. Structure/function relationships of axon-associated adhesion receptors of the immunoglobulin superfamily. Curr Opin Neurobiol 1996; 6, 584-593.

6. Grant SG, O'Dell TJ. Multiprotein complex signaling and the plasticity problem.Curr Opin Neurobiol 2001 ;11(3):363-368.
7. Kenwrick S, Doherty P. Neural cell adhesion molecule L1: relating disease to function.Bioessays. 1998;20(8):668-675.
8. Kamiguchi H, Hlavin ML, Lemmon V. Role of L1 in neural development: what the knock-outs tell us. Mol Cell Neurosci 1998;12(1-2):48-55.
9. Kenwrick S, Watkins A, De Angelis E. Neural cell recognition molecule L1: relating biological complexity to human disease mutations.Hum Mol Genet 2000 12;9(6):879-886.
10. Dahme M, Bartsch U, Martini R et al. Disruption of the mouse L1 gene leads to malformations of the nervous system. Nat Genet 1997 17, 346-349.
11. Cohen NR, Taylor JS, Scott LB et al. Errors in corticospinal axon guidance in mice lacking the neural cell adhesion molecule L1. Curr Biol 1998 8, 26-33.
12. Stanfield BB. The development of the corticospinal projection. Prog Neurobiol 1992;38(2):169-202.
13. Joosten EA, Bar DP. Axon guidance of outgrowing corticospinal fibres in the rat.J Anat 1999;194 (Pt 1):15-32.
14. Castellani V, Chédotal A, Schachner M et al. Analysis of L1-deficient mouse phenotype reveals a cross-talk between Sema3A and L1 signaling pathways in axonal guidance. Neuron 2000; 27, 237-249.
15. Culotti JG, Merz DC. DCC and netrins. Curr Opin Çell Biol 1998;10(5):609-613.
16. Guthrie S. Axon guidance: Robos make the rules. Curr Biol 2001;17;11(8):R300-3.
17. Masuda T, Okado N, Shiga T. The involvement of axonin-1/SC2 in mediating notochord-derived chemorepulsive activities for dorsal root ganglion neurites. Dev Biol. 2000; 15;224(2):112-121.
18. Song H, Ming G, He Z et al. Conversion of neuronal growth cone responses from repulsion to attraction by cyclic nucleotides. Science 1998; 281, 1515-8.
19. Nayeem N. Silletti S, Yang X et al. A potential role for the plasmin(ogen) system in the posttranslational cleavage of the neural cell adhesion molecule L1. *J Cell* Sci 1999; 112, 4739-4749.
20. Gutwein P, Oleszewski M, Mechtersheimer S et al. Role of Src kinases in the ADAM-mediated release of L1 adhesion molecule from human tumor cells. J Biol Chem 2000; 19;275(20):15490-7
21. Mechtersheimer S, Gutwein P, Agmon-Levin N et al. Ectodomain shedding of L1 adhesion molecule promotes cell migration by autocrine binding to integrins. J Cell Biol 2001;12,661-674.
22. Dodd J, Jessell TM. Axon guidance and the patterning of neuronal projections in vertebrates.Science 1988; 4;242(4879):692-9.
23. Bolz J, Kossel A, Bagnard D. The specificity of interactions between the cortex and the thalamus. Ciba Found Symp. 1995;193:173-91; discussion 192-9.
24. Bagnard D, Chounlamountri N, Puschel AW et al. Axonal surface molecules act in combination with semaphorin 3a during the establishment of corticothalamic projections. Cereb Cortex 2001;11(3):278-285.
25. Jacob J, Hacker A, Guthrie S. Mechanisms and molecules in motor neuron specification and axon pathfinding.Bioessays 2001;23(7):582-595.
26. Fujisawa H, Kitsukawa T, Kawakami A et al. Roles of a neuronal cell-surface molecule, neuropilin, in nerve fiber fasciculation and guidance. Cell Tissue Res 1997;290(2):465-470.
27. Kawakami A, Kitsukawa T, Takagi S et al. Developmentally regulated expression of a cell surface protein, neuropilin, in the mouse nervous system. J Neurobiol 1996;29(1):1-17.
28. Kitsukawa T, Shimizu M, Sanbo M et al. Neuropilin-semaphorin III/D-mediated chemorepulsive signals play a crucial role in peripheral nerve projection in mice. Neuron 1997;19, 995-1005.
29. Stoeckli ET, Landmesser LT. Axon guidance at choice points. Curr Opin Neurobiol 1998;8(1):73-79.

30. Kaprielian Z, Imondi R, Runko E. Axon guidance at the midline of the developing CNS. Anat Rec 2000;15;261(5):176-197.
31. Song HJ, Poo MM. Signal transduction underlying growth cone guidance by diffusible factors.Curr Opin Neurobiol 1999;9(3):355-363.
32. Dickson BJ. Rho GTPases in growth cone guidance. Curr Opin Neurobiol 2001;11(1):103-110.
33. Stein E, Tessier-Lavigne M. Hierarchical organization of guidance receptors: silencing of netrin attraction by slit through a Robo/DCC receptor complex. Science 2001;9;291(5510):1928-1938.

NEUROPILIN AND ITS LIGANDS IN NORMAL LUNG AND CANCER

Joëlle Roche[1], Harry Drabkin[2] and Elisabeth Brambilla[3]

SUMMARY

Neuropilins (NRPs) are receptors for class 3 Semaphorins and function as co-receptors for Vascular endothelial growth factor isoforms, $VEGF_{165}$ and $VEGF_{145}$ and related molecules. NRPs are expressed in a variety of neural and non-neural tissues and are required for normal development. Interestingly, class 3 Semaphorins and VEGF compete for common NRP binding. As a consequence, Semaphorins and VEGF appear to be mutually antagonistic. In the lung, NRP levels increase during development and NRPs and Semaphorins are involved in lung branching, probably by altering cell morphology or by regulating cell motility and migration. During lung tumorigenesis, both NRP and VEGF expression increase on dysplastic lung epithelial cells; SEMA3F expression is reduced and SEMA3F protein is delocalized from the membrane to the cytoplasm. In lung cancers, SEMA3F staining correlates inversely with tumor stage with high SEMA3F associated with less aggressive tumors. Conversely, more aggressive tumors are associated with increased VEGF staining and a corresponding loss in membranous SEMA3F.

INTRODUCTION

Neuropilin (NRP) 1 and 2 are transmembrane glycoproteins involved in neuronal cell guidance, axon growth and fasciculation.[1] In addition to its role in the nervous system, NRP1 is expressed in the developing heart, vasculature, skeleton and lung. NRP2 has a similar expression profile. Neuropilins are receptors for two

[1]Université de Poitiers, 40 Av du Recteur Pineau, 86022 Poitiers Cédex France. [2]University of Colorado Health Sciences Center, Division of Medical Oncology, Box B171, 4200 East Ninth Avenue, Denver, CO 80262, USA [3]Laboratoire de Pathologie Cellulaire, INSERM EMI, CHRU Grenoble, 38043 Grenoble Cédex 09, France.

types of very different ligands: semaphorins[2,3] and vascular endothelial growth factor, VEGF.[4]

Semaphorins are a large family of secreted and membrane associated molecules containing a characteristic 500 amino acid Sema domain. They have been classified into eight groups based on their overall similarity and structural features.[5] Collapsin, now known as Sema3A, was originally identified on the basis of its chemorepellent activity. Secreted semaphorins from class 3 are the only semaphorins that bind neuropilins and have been implicated in axon steering, fasciculation, branching and synapse formation.[6] While Sema3A only binds NRP1, Sema3C binds NRP1 and NRP2 equally whereas Sema3F has greater affinity for NRP2 than NRP1.[7] This binding is essential for semaphorin function[2,3,8] and NRP2 is the functional receptor for Sema3F in the nervous system.[9-12] Other molecules are necessary to transduce semaphorin signals which include plexins[13] and collapse response mediator protein CRMP.[14]

VEGF, a 40-45 kDa homodimeric protein, regulates normal embryonic vasculogenesis, physiological angiogenesis and tumor angiogenesis. Originally defined as an endothelial cell (EC) mitogen and chemotactic factor, there is now growing evidence that VEGF stimulates non-EC cells.[15-18] Five different isoforms of VEGF monomers consisting of 121, 145, 165, 186 and 206 amino acids produced by alternative splicing have been identified with $VEGF_{121}$ and $VEGF_{165}$ being the most abundant.

NRP1 was identified as a receptor for $VEGF_{165}$ but not for $VEGF_{121}$.[4] NRP2 binds both $VEGF_{165}$ and $VEGF_{145}$.[19] Importantly, the presence of NRP1 together with the high affinity VEGF receptor 2 (KDR/flk-1) result in greater tyrosine kinase activity. Further support for the role NRPs in cardiovascular development comes from studies utilizing transgenic and knock-out mice. Mice that overexpress NRP1 develop excess capillaries and blood vessels, dilatation of blood vessels and heart defects in addition to neurological abnormalities.[20] When deleted for NRP1, mutant mice die during the second half of gestation. In addition to neurological defects, NRP1 $^{-/-}$ mice exhibit severe defects in the cardiovascular system reflecting either a requirement for Semaphorin signaling and / or the presence of NRP1 as a receptor for critical VEGF isoforms.[8] Disruption of Sema3A also causes severe abnormalities in neural and non-neural tissues including hypertrophy of the right ventricle and dilatation of the right atrium.[21] The discovery that Neuropilins were capable of binding two distinct ligands suggested that class 3 Semaphorins and VEGF might compete. A competitive interaction was documented between Sema3A and VEGF in endothelial cells.[22,23]

In addition to their role in the nervous system and in angiogenesis, Neuropilins and Semaphorins have been implicated in other developmental processes and in tumorigenesis. For the remainder of this discussion, we will focus on these molecules in the development of normal lung and lung cancer.

NEUROPILIN AND SEMAPHORIN IN NORMAL MICE LUNG DEVELOPMENT

Fetal lung development involves coordinated cell proliferation, migration, branching morphogenesis and differentiation and normal development depends on reciprocal induction between epithelial and mesenchymal cells. Several growth factors are known to affect lung epithelial cell proliferation. For example, epidermal growth factor and fibroblast growth factors positively influence proliferation, whereas bone morphogenetic protein BMP-4 and TGFβ have negative effects.[24-28] Many factors which affect branching morphogenesis are also becoming elucidated. Guidance molecules such as Semaphorins are likely candidates to affect these processes as both neuropilins and semaphorins are expressed in the lung.[2,3,11,21,29,30] Moreover, rCRMP-2 which is an intra-cellular protein required for Sema3A signaling[14] is expressed in the lung[31] as it is also the case for CRMP-1.[32]

Only a few lung effects have been reported in mice overexpressing NRP1 or in mice with a NRP1 knock-out. When the lung was examined in NRP1 homozygous null ($nrp1^{-/-}$) animals, it was found to be smaller and the number of branches in the left lung significantly lower than in wild-type ($nrp1^{+/+}$) or heterozygous ($nrp1^{+/-}$) animals.[33] In contrast to $nrp1^{-/-}$ mice, many $nrp2^{-/-}$ mice survive into adulthood despite the existence of numerous neurological deficits, some of which are complementary to those observed previously in NRP1 mutants.[34,35] However, no effects involving the lungs were reported for either the $nrp2^{-/-}$ mice or for Sema3A knockouts.[21,36,37]

The absence of severe developmental effects in the lung may be the result of redundancy. Other data indicate that semaphorins and neuropilins are likely to be critically involved in lung development.[33,38] Expression studies indicate that Sema3A is expressed at high levels mainly in the distal mesenchyme around the airway epithelium and this expression decreases with time. In epithelial cells, Sema3A is strongly expressed in intermediate bronchioles at E15.5. Expression of Sema3C was restricted predominantly to the lobar bronchus. These expression patterns overlapped or were adjacent to regions expressing NRP1 and NRP2. In contrast, SEMA3F expression was weak and diffuse in the distal epithelium and in the surrounding mesenchyme in early stages, but became confined to the terminal epithelium by E15.5. NRP1 levels rise dramatically during development along with CRMP-2 immunoreactivity in developing and adult alveolar epithelium.

Ito et al.[33] treated lung explant culture with different semaphorins. A striking feature of early lung development is the budding and branching which is retained even in culture. Treatment of explant cultures with Sema3A resulted in fewer terminal buds. Co-treatment with a soluble form of NRP1[39] lacking the transmembrane and intracellular regions attenuated the Sema3A effects whereas sNRP1 alone had no activity. The reduction in terminal buds was not attributable to growth as no effects were detected with BrdU incorporation. In contrast, Sema3C and Sema3F stimulated branching morphogenesis in lung explants from fetal mice and these effects were blocked with sNRP1 or sNRP2, respectively.[38] Cell proliferation was stimulated as shown by BrdU labeling. These data indicate that multiple semaphorins exert counterbalancing effects on branching morphogenesis, constituting a novel regulatory system in lung development. Dual effects of Semaphorins were first reported in the nervous system[10,12,40] which, in at least some cases, are the result of different cGMP concentrations.[41]

How do Semaphorins/Neuropilins affect lung branching? One hypothesis from Kagoshima et al.[38] is that Semaphorins could promote or inhibit airway branching by the alteration of cell morphology or by the regulation of cell motility and migration. Alterations of cell morphology were seen in COS cells transfected with Sema3A[42] and in mammary adenocarcinoma cells transfected with SEMA3F (Nasarre et al., submitted). In the later case, transfected cells rounded up and detached. Cell motility and migration might also be involved as Sema3A inhibits endothelial cell motility[22] and has been shown to regulate neural crest migration.[43] Likewise, in *C. elegans*, Semaphorin-2a prevents ectopic cell contacts during epidermal morphogenesis.[44] We also demonstrated that SEMA3F is localized in motile regions such as in leading edges or ruffling membranes of lamellipodia in HeLa cells.[45]

NEUROPILINS AND ITS LIGANDS IN HUMAN LUNG TUMOR

Following the cloning of SEMA3F, by our group and others, from a recurrent homozygous deletion region in small-cell lung cancer (SCLC)[46-48] and SEMA3B[48], we were intrigued by the possibility that neural guidance molecules such as semaphorins could be involved in lung tumorigenesis. This chromosomal region is well known for loss of heterozygosity (LOH) as an early event in lung tumors and was postulated to contain a tumor suppressor gene.[49,50] More direct evidence for such an activity came from the transfection of P1 clones containing SEMA3F into a mouse tumor cell line[51] and it was also shown that SEMA3F by itself suppresses tumor formation in nude mice.[52] It is also notable that another 3p homozygous deletion region, identified in the SCLC cell line U2020 encodes a repulsive neural guidance molecule DUTT1 (Deleted in U Twenty-Twenty).[53] DUTT1 is the probable human homologue of the *Drosophila* gene Roundabout (Robo)[54] which is the receptor for the midline ligand, Slit.

Several reports in the literature have implicated semaphorins in cancers as survival factors with increased metastatic ability. SEMA3E was identified in human cancer to confer non-MDR (Multi Drug Resistance) resistance[55] and was also overexpressed in metastatic human lung adenocarcinomas.[56] Similarly, Sema3E expression has been correlated with the metastatic ability of certains tumors.[57] SEMA4D (CD100) downregulation occurs in non-Hodgkin's B-cell lymphomas and has been postulated to regulate adhesiveness and metastatic potential.[58] Therefore some semaphorins show overexpression in tumors whereas others are downregulated. This may reflect the bifunctional effects of semaphorins previously observed in the nervous system.

In normal lung, we studied SEMA3F expression using a specific affinity purified antibody.[59] SEMA3F expression was found in epithelial cells. In large bronchi, there was strong membrane staining in addition to mild diffuse cytoplasmic staining.[45] In bronchioles, SEMA3F was restricted to basal epithelial cells. Endothelial cells of the alveolar capillary bed did not express SEMA3F, whereas about 20% of vessels more than 100 mm diameter were positive for expression. In lung tumors, SEMA3F localization was predominantly cytoplasmic and the overall levels were reduced (Fig. 1). In resected NSCLC cancers (Non Small Cell Lung Cancer), low levels of SEMA3F correlated with higher stage (more aggressive) disease. In all lung cancer subtypes, an exclusive cytoplasmic localization of SEMA3F was associated with VEGF overexpression which suggested that SEMA3F could compete with VEGF for binding to cell surface NRP receptors.[45] These studies have now been expanded to include 112 lung cancers and 50 preneoplasic lesions (Lantuejoul et al., submitted). In preneoplasic lesions, SEMA3F was low indicating that loss of SEMA3F protein, like the previous LOH studies would predict, is an early event in lung tumorigenicity. Recently, expression of CRMP-1, a mediator in the Semaphorin pathway, was found to be inversely correlated with the invasive capability of lung cancer cell lines.[32] In normal lung, we found NRP1 and NRP2 expressed in bronchial basal cells (Lantuejoul et al., submitted). In preinvasive bronchial lesions NRP1 and NRP2 expression was significantly increased from hyperplastic mucosa to moderate dysplasia with a plateau reached in severe dysplasia (Fig. 1). Increased neuropilin staining was also observed in conjunction with increased VEGF.

Interestingly, we observed using a wound assay of HeLa cells that cells at the border of the wound had increased staining for NRP1 but NRP1 was translocated to the cytoplasm (Fig. 2). Since cells at the wound border are apparently stimulated to migrate, up-regulation of NRP1 and translocation to the cytoplasm would be expected to facilitate this process (Lantuejoul et al., submitted). NRP1 has been previously implicated in tumor progression through its effects on angiogenesis and NRP1 overexpression likely represents a biomarker for tumor aggressiveness. In prostate carcinoma AT2.1 cells, overexpression of NRP1 resulted in increased basal cell motility and $VEGF_{165}$ binding.[60] Furthermore, the tumors were enlarged in vivo and showed increased microvessel density, proliferation of endothelial cells, dilated blood vessels and, notably, less tumor cell apoptosis.[60] The expression of NRP1 in Neuropilin-deficient breast carcinoma cells protects them from apoptosis.[61] NRP1 expression has also been correlated with an advanced stage of prostatic cancer and malignant behavior in astrocytomas.[62,63] Likewise, NRP1 was higher in rat estrogen-induced pituitary tumors and promoted angiogenesis.[64] In addition, experimentally overexpressed soluble NRP1 (sNRP1), a naturally occurring antagonist, leads to tumors which are apoptotic, hemorrhagic and full of disrupted blood vessels.[65]

VEGF is expressed in normal lung by bronchial basal cells as well as hyperplastic type II pneumocytes in addition to endothelial cells. Expression includes a frequent reinforcement at the membrane. We found that VEGF increased significantly with the histological grades of preneoplasic lesions and culminated in corresponding invasive carcinoma in parallel to neuropilins (Fig. 1). Lung tumors stained

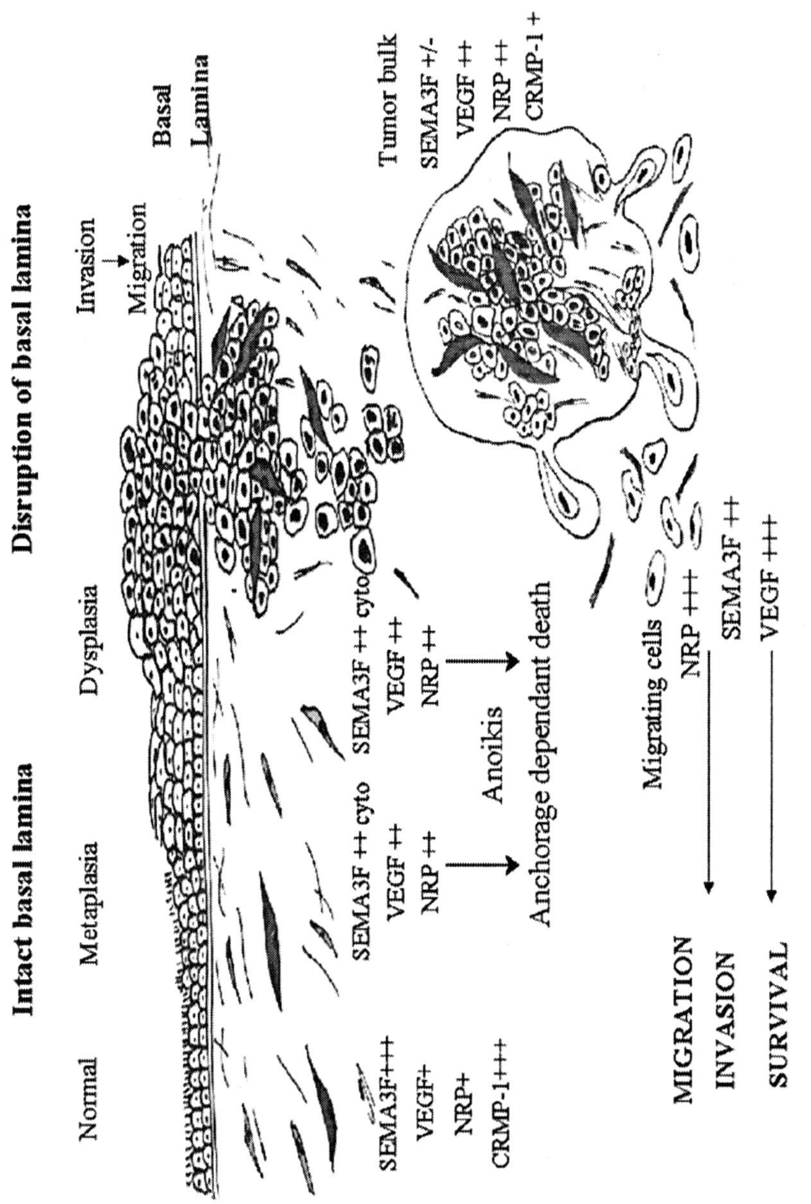

Figure 1. SEMA3F, VEGF, Neuropilins and CRMP-1 levels during lung tumor progression. SEMA3F is present in normal lung but its level decreases in preneoplasic lesions indicating that loss of SEMA3F is an early event. In addition, SEMA3F is found in the cytoplasm instead at the membrane. From the results of Shi et al.[31] on cell lines, CRMP-1 is inversely correlated with the invasive capabilility and would decrease from normal lung to tumor. In contrast, NRP1 and VEGF levels increases during tumor lung progression. In metaplasia and dysplasia, anoikis and anchorage dependant death would take place. Isolated clusters of tumor cells in lung tumors stain strongly for SEMA3F, Neuropilins and VEGF leading probably to migration, invasion and survival.

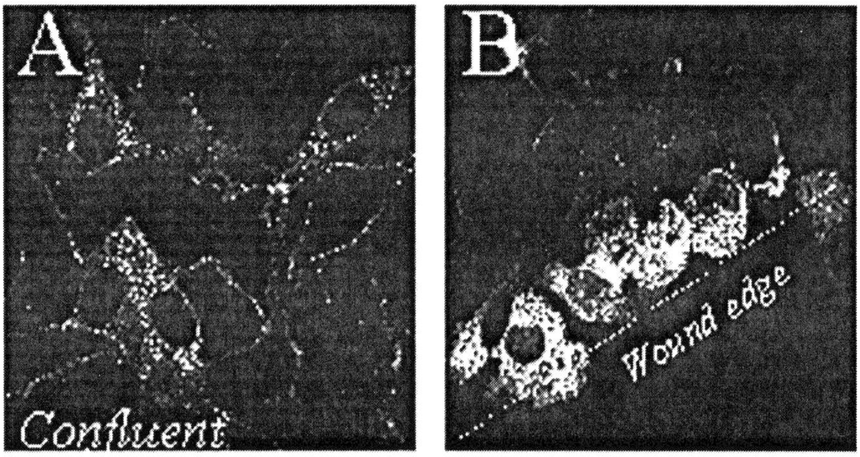

Figure 2. NRP1 staining at the border of a wound in a confluent HeLa cell culture. Cells were immunostained with a polyclonal anti-NRP1 antibody provided by A Kolodkin (dilution 1/1000). Only some cells are positive in a confluent culture (A) but cells at the border of the wound demonstrate increased positivity (B). The border is delineated.

positively for VEGF with more intense staining at the periphery of tumor lobules than inside. VEGF may stimulate an autocrine signaling pathway, independent of angiogenesis, to maintain cell survival as it was proposed for NRP1 expressing breast carcinoma cell lines.[61] Surprisingly, we also found isolated clusters of tumors cells in lung tumors which stained strongly for SEMA3F, Neuropilins and VEGF (Fig. 3). While as yet unproven, this raises the possibility that Semaphorin expression may be dynamic as has been reported for β-catenin and E-cadherin in colon tumors[66] and in breast cancers undergoing migration in vitro.[67] However, overexpression of SEMA3B in lung carcinoma cell lines induces apoptosis[68] and Semaphorins were also described as death inducers in sensory neurons[69] and neural progenitors.[23] Therefore, extra SEMA3F would rather lead to elimination of transformed cells but there is a balance between SEMA3F and VEGF and it is hard to predict on which side, proliferation or apoptosis, cells will go.

A model for these various interactions is shown in Figure 4. Normal stationary cells in the lung express Neuropilins and a substantial amount of SEMA3F. During the process of tumor development, VEGF and Neuropilin expression increase while SEMA3F binding to the surface of epithelial cells declines. Further downregulation of SEMA3F occurs at the transcriptional level. It is possible that hypoxia may regulate components of the system other than VEGF but this has not been reported. Not only does the reduction in SEMA3F levels facilitate growth or survival activities of VEGF on primary tumors, which appears to occur even in the absence of VEGFR2 (Vascular endothelial growth factor Receptor 2/ KDR/ Flk-1), but increased VEGF levels compete for Semaphorin binding and overcome its inhibitory actions. With this scenario, we would anticipate that semaphorin replacement combined with anti-VEGF

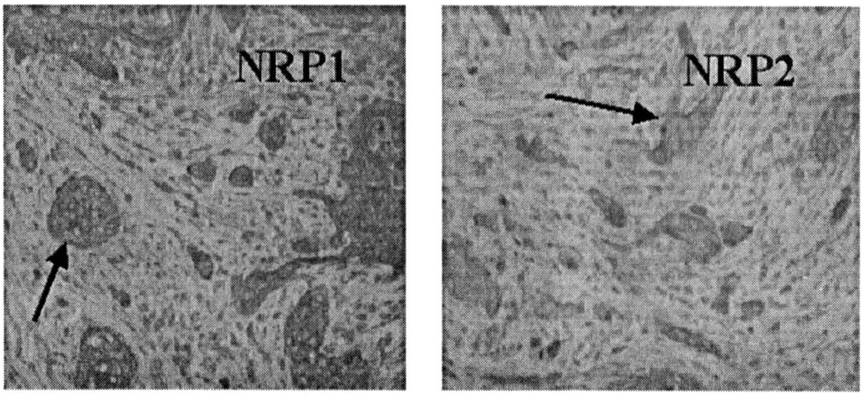

Figure 3. Squamous cell carcinoma with small clusters of cell isolated in the stroma, evading from the tumor bulk on serial sections. Samples were incubated with rabbit polyclonal NRP1 and NRP2 antibodies from Santa Cruz Biotechnology (Santa Cruz, CA USA) at 1/100 dilution. Immunoreactivity was detected by peroxidase activity. Cells are strongly positive for NRP1 in the cytoplams with membrane reinforcement (original magnification x200) and show a strong cytoplasmic staining for NRP2 (original magnification x100).

therapies should be additive or even synergistic in the treatment of established tumors or preneoplastic lesions.

ACKNOWLEDDGEMENTS

This work was supported by CNRS, ARC and Ligue Nationale Contre le Cancer for JR, by the University of Colorado Lung Cancer SPORE CA5187-07 for HD and by INSERM, Ligure Nationale Contre le Cancer and PHRC 1999 for EB.

REFERENCES

1. Fujisawa H, Kitsukawa T. Receptors for collapsin/Semaphorins. Curr. Opin. Neurobiol. 1998; 8:587-592
2. He Z, Tessier-Lavigne M. Neuropilin is a receptor for the axonal chemorepellent Semaphorin III. Cell 1997; 90:739-751
3. Kolodkin A, Levengood D, Rowe E et al. Neuropilin is a Semaphorin III receptor. Cell 1997; 90:753-762
4. Soker S, Takashima S, Miao H et al. Neuropilin-1 is expressed by endothelial and tumor cells as an isoform-specific receptor for vascular endothelial growth factor. Cell 1998; 92:735-745
5. Unified nomenclature for the Semaphorins/collapsins. Semaphorin Nomenclature Committee. Cell 1999; 97:551-2
6. Raper J. Semaphorins and their receptors in vertebrates and invertebrates. Current Opinion in Neurobiology 2000; 10:88-94.
7. Chen H, Chédotal A, He Z et al. Neuropilin-2, a novel member of the Neuropilin family, is a high affinity receptor for the Semaphorins Sema E and Sema IV but not Sema III. Neuron 1997; 19:547-559.

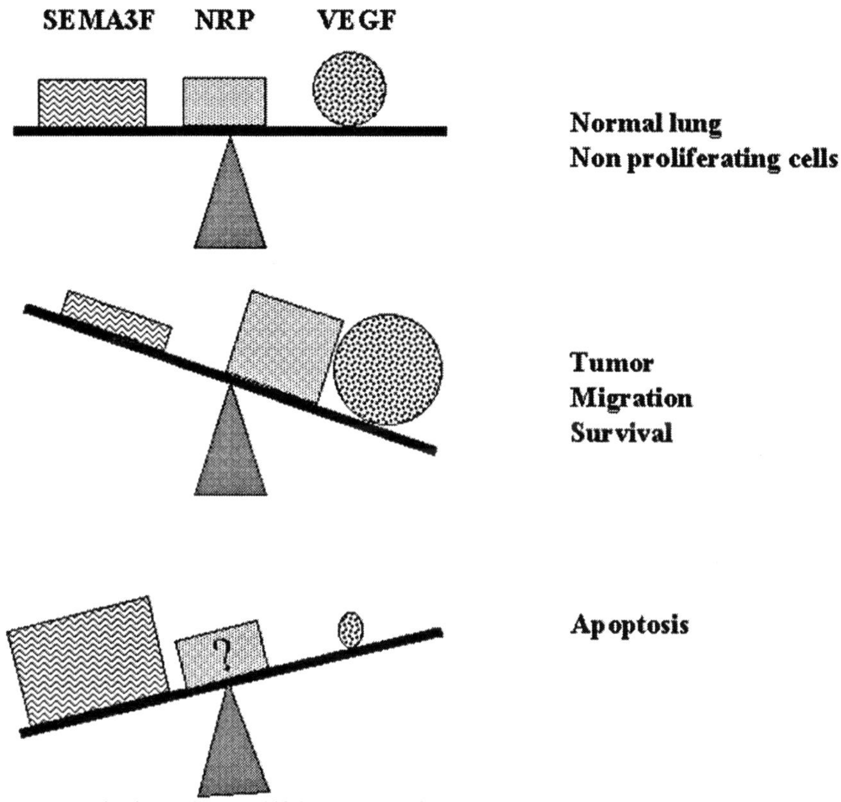

Figure 4. Model of how cells may shift from non proliferation to tumorigenecity or apoptosis. Increasing VEGF and NRP levels along with decreases in SEMA3F promote tumorigenesis, migration and survival. In contrast, a high level of SEMA3F with a low level of VEGF would lead to apoptosis. The status of NRP is not known in apoptotic cells.

8. Kitsukawa T, Shimizu M, Sanbo M et al. Neuropilin-Semaphorin III/D-mediated chemorepulsive signals play a crucial role in peripheral nerve projection in mice. Neuron 1997; 19:995-1005.
9. Chedotal A, Del Rio JA, Ruiz M et al. Semaphorins III and IV repel hippocampal axons via two distinct receptors. Development 1998; 125:4313-4323.
10. Chen H, He Z, Tessier-Lavigne M. Axon guidance mechanisms: Semaphorins as simultaneous repellents and anti-repellents. Nat. Neurosci. 1998; 1:436-439.
11. Giger RJ, Urquhart ER, Gillespie SK et al. Neuropilin-2 is a receptor for Semaphorin IV: insight into the structural basis of receptor function and specificity. Neuron 1998; 21:1079-1092.
12. Takahashi T, Nakamura F, Jin Z et al. Semaphorins A and E act as antagonists of Neuropilin-1 and agonists of Neuropilin-2 receptors. Nat. Neurosci. 1998; 1:487-493.
13. Tamagnone L, Artigiani S, Chen H et al. Plexins are a large family of receptors for transmembrane, secreted, and GPI-anchored Semaphorins in vertebrates. Cell 1999; 99:71-80.
14. Goshima Y, Nakamura F, Strittmatter P et al. Collapsin-induced growth cone collapse mediated by an intracellular protein related to UNC-33. Nature 1995; 376:509-514.

15. Deckers MM, Karperien M, van der Bent C et al. Expression of vascular endothelial growth factors and their receptors during osteoblast differentiation. Endocrinology 2000; 141:1667-74.
16. Midy V, Plouet J. Vasculotropin/vascular endothelial growth factor induces differentiation in cultured osteoblasts. Biochem Biophys Res Commun 1994; 199:380-6.
17. Sondell M, Lundborg G, Kanje M. Vascular endothelial growth factor stimulates Schwann cell invasion and neovascularization of acellular nerve grafts. Brain Res 1999; 846:219-28.
18. Byzova TV, Goldman CK, Pampori N et al. A mechanism for modulation of cellular responses to VEGF: activation of the integrins. Mol Cell 2000; 6:851-60.
19. Gluzman-Poltorak Z, Cohen T, Herzog Y et al. Neuropilin-2 and Neuropilin-1 are receptors for $VEGF_{165}$ and PLGF- 2, but only Neuropilin-2 functions as a receptor for $VEGF_{145}$. J Biol Chem 2000; 275:18040-18045.
20. Kitsukawa T, Shimono A, Kawakami A et al. Overexpression of a membrane protein, Neuropilin, in chimeric mice causes anomalies in the cardiovascular system, nervous system and limbs. Development 1995; 121:4309-18.
21. Behar O, Golden JA, Mashimo H et al. Semaphorin III is needed for normal patterning and growth of nerves, bones and heart. Nature 1996; 383:525-528.
22. Miao HQ, Soker S, Feiner L et al. Neuropilin-1 mediates collapsin-1/Semaphorin III inhibition of endothelial cell motility: functional competition of collapsin-1 and vascular endothelial growth factor-165. J Cell Biol 1999; 146:233-42.
23. Bagnard D, Vaillant C, Khuth ST et al. Semaphorin 3A-vascular endothelial growth factor-165 balance mediates migration and apoptosis of neural progenitor cells by the recruitment of shared receptor. J Neurosci 2001; 21:3332-41.
24. Hogan BL, Yingling JM. Epithelial/mesenchymal interactions and branching morphogenesis of the lung. Curr Opin Genet Dev 1998; 8:481-6.
25. Metzger RJ, Krasnow MA. Genetic control of branching morphogenesis. Science 1999; 284:1635-9.
26. Cardoso WV. Lung morphogenesis revisited: old facts, current ideas. Dev Dyn 2000; 219:121-30.
27. Warburton D, Schwarz M, Tefft D et al. The molecular basis of lung morphogenesis. Mech Dev 2000; 92:55-81.
28. Bellusci S, de Maximy A, Thiéry JP. Contrôle moléculaire de la morphogénèse pulmonaire chez la souris. Médecine Sciences 1999; 15:815-822.
29. Luo Y, Raible D, Raper A. Collapsin: a protein in brain that induces the collapse and paralysis of neuronal growth cones. Cell 1993; 75:217-227.
30. Takahashi T, Nakamura F, Stittmatter S. Neuronal and non-neuronal collapsin-1 binding sites in developing chick are distinct from other Semaphorin binding sites. The Journal of Neuroscience 1997; 17:9183-9193.
31. Wang L, Strittmatter S. A family of rat CRMP genes is differentially expressed in the nervous system. J Neurosci 1996; 16:6197-6207.
32. Shih JY, Yang SC, Hong TM et al. Collapsin response mediator protein-1 and the invasion and metastasis of cancer cells. J Nat Cancer Inst 2001, 93, 1392-1400.
33. Ito T, Kagoshima M, Sasaki Y et al. Repulsive axon guidance molecule Sema3A inhibits branching morphogenesis of fetal mouse lung. Mech Dev 2000; 97:35-45.
34. Giger RJ, Cloutier JF, Sahay A et al. Neuropilin-2 is required in vivo for selective axon guidance responses to secreted Semaphorins. Neuron 2000; 25:29-41.
35. Chen H, Bagri A, Zupicich JA et al. Neuropilin-2 regulates the development of selective cranial and sensory nerves and hippocampal mossy fiber projections. Neuron 2000; 25:43-56
36. Taniguchi M, Yuasa S, Fujisawa H et al. Disruption of Semaphorin III/D gene causes severe abnormality in peripheral nerve projection. Neuron 1997; 19:519-530.
37. White FA, Behar O. The development and subsequent elimination of aberrant peripheral axon projections in Semaphorin3A null mutant mice. Dev Biol 2000; 225:79-86.

38. Kagoshima M, Ito T. Diverse gene expression and function of Semaphorins in developing lung: positive and negative regulatory roles of Semaphorins in lung branching morphogenesis. Genes Cells 2001; 6:559-71.
39. Goshima Y, Hori H, Sasaki Y et al. Growth cone Neuropilin-1 mediates collapsin-1/Sema III facilitation of antero- and retrograde axoplasmic transport. J Neurobiol 1999; 39:579-89
40. Bagnard D, Lohrum M, Uziel D et al. Semaphorins act as attractive and repulsive guidance signals during the development of cortical projections. Development 1998; 125:5043-5053
41. Song H, Ming G, He Z et al. Conversion of neuronal growth cone responses from repulsion to attraction by cyclic nucleotides. Science 1998; 281:1515-1518.
42. Takahashi T, Fournier A, Nakamura F et al. Plexin-Neuropilin-1 complexes form functional Semaphorin-3A receptors. Cell 1999; 99:59-69.
43. Eickholt B, Mackenzie S, Graham A et al. Evidence for collapsin-1 functioning in the control of neural crest migration in both trunk and hindbrain regions. Development 1999; 126:2181-2189.
44. Roy PJ, Zheng H, Warren CE et al. mab-20 encodes Semaphorin-2a and is required to prevent ectopic cell contacts during epidermal morphogenesis in Caenorhabditis elegans. Development 2000; 127:755-767.
45. Brambilla E, Constantin B, Drabkin H et al. Semaphorin SEMA3F localization in malignant human lung and cell lines : A suggested role in cell adhesion and cell migration. Am J Pathol 2000; 156:939-950.
46. Roche J, Boldog F, Robinson M et al. Distinct 3p21.3 deletions in lung cancer, analysis of deleted genes and identification of a new human Semaphorin. Oncogene 1996; 12:1289-1297.
47. Xiang R, Hensel C, Garcia D et al. Isolation of the human Semaphorin III/F gene (SEMA3F) at chromosome 3p21, a region deleted in lung cancer. Genomics 1996; 32:39-48.
48. Sekido Y, Bader S, Latif F et al. Human Semaphorins A (V) and (IV) reside in the 3p21.3 small cell lung cancer deletion region and demonstrate distinct expression patterns. Proc Natl Acad Sci, USA 1996; 93:4120-4125.
49. Kok K, Naylor S, Buys C. Deletions of the short arm of chromosome 3 in solid tumors and the search for suppressor genes. Adv Cancer Res 1997; 71:27-92.
50. Lerman MI, Minna JD. The 630-kb lung cancer homozygous deletion region on human chromosome 3p21.3: identification and evaluation of the resident candidate tumor suppressor genes. The International Lung Cancer Chromosome 3p21.3 Tumor Suppressor Gene Consortium. Cancer Res 2000; 60:6116-33.
51. Todd M, Xiang R, Garcia D et al. An 80 Kb P1 clone from chromosome 3p21.3 suppresses tumor growth in vivo. Oncogene 1996; 13:2387-2396.
52. Xiang R, Xhou X, Tse C et al. Expression of human Semaphorin 3F in a human ovarian cancer cell line (Hey) suppresses tumor formation in nude mice and blocks program cell death caused by adriamycin or taxol. 91st Proceedings of the American Association for Cancer Research, Abstract 5216, San Francisco 2000.
53. Sundaresan V, Roberts I, Bateman A et al. The DUTT1 Gene, a Novel NCAM Family Member Is Expressed in Developing Murine Neural Tissues and Has an Unusually Broad Pattern of Expression. Mol Cell Neurosci 1998; 11:29-35.
54. Kidd T, Brose K, Mitchell KJ et al. Roundabout controls axon crossing of the CNS midline and defines a novel subfamily of evolutionarily conserved guidance receptors. Cell 1998; 92:205-15.
55. Yamada T, Endo R, Gotoh M et al. Identification of Semaphorin E as non-MDR drug resistance gene of human cancers. Proc. Natl. Acad. Sci., USA 1997; 94:14713-14718.
56. Martin-Satue M, Blanco J. Identification of Semaphorin E gene expression in metastatic human lung adenocarcinoma cells by mRNA differential display. J Surg Oncol 1999; 72:18-23
57. Christensen CR, Klingelhofer J, Tarabykina S et al. Transcription of a novel mouse Semaphorin gene, M-semaH, correlates with the metastatic ability of mouse tumor cell lines. Cancer Res 1998; 58:1238-1244.

58. Dorfman D, Shahsafaei A, Nadler L et al. The leukocyte Semaphorin CD100 is expressed in most T-cell, but few B-cell, non-Hodgkin's lymphomas. Am. J. Pathol. 1998; 153:255-262
59. Hirsch E, Hu L-J, Prigent A et al. Distribution of Semaphorin IV in adult human brain. Brain Res. 1999; 823:67-79.
60. Miao HQ, Lee P, Lin H et al. Neuropilin-1 expression by tumor cells promotes tumor angiogenesis and progression. Faseb J 2000; 14:2532-9.
61. Bachelder RE, Crago A, Chung J et al. Vascular endothelial growth factor is an autocrine survival factor for Neuropilin-expressing breast carcinoma cells. Cancer Res 2001; 61:5736-40.
62. Latil A, Bieche I, Pesche S et al. VEGF overexpression in clinically localized prostate tumors and Neuropilin-1 overexpression in metastatic forms. Int J Cancer 2000; 89:167-71
63. Ding H, Wu X, Roncari L et al. Expression and regulation of Neuropilin-1 in human astrocytomas. Int J Cancer 2000; 88:584-92.
64. Banerjee SK, Zoubine MN, Tran TM et al. Overexpression of vascular endothelial growth factor164 and its co- receptor Neuropilin-1 in estrogen-induced rat pituitary tumors and GH3 rat pituitary tumor cells. Int J Oncol 2000; 16:253-60.
65. Gagnon ML, Bielenberg DR, Gechtman Z et al. Identification of a natural soluble Neuropilin-1 that binds vascular endothelial growth factor: In vivo expression and antitumor activity. Proc Natl Acad Sci USA 2000; 97:2573-8.
66. Brabletz T, Jung A, Hermann K et al. Nuclear overexpression of the oncoprotein beta-catenin in colorectal cancer is localized predominantly at the invasion front. Pathol Res Pract 1998; 194:701-4.
67. Graff JR, Gabrielson E, Fujii H et al. Methylation patterns of the E-cadherin 5' CpG island are unstable and reflect the dynamic, heterogeneous loss of E-cadherin expression during metastatic progression. J Biol Chem 2000; 275:2727-32.
68. Tomizawa Y, Sekido Y, Kondo M. et al. Inhibition of lung cancer cell growth and induction of apoptosis following re-expression of 3p21.3 tumor suppressor gene candidate SEMA3B. Proc Natl Acad of Sci USA 2001; 98:13954-13959.
69. Gagliardini V, Fankhauser C. Semaphorin III Can Induce Death in Sensory Neurons. Mol Cell Neurosci 1999; 14:301-316.

NEUROPILIN AND CLASS 3 SEMAPHORINS IN NERVOUS SYSTEM REGENERATION

Fred De Winter, Anthony J.G.D. Holtmaat and Joost Verhaagen

SUMMARY

Injury to the mature mammalian central nervous system (CNS) is often accompanied by permanent loss of function of the damaged neural circuits. The failure of injured CNS axons to regenerate is thought to be caused, in part, by neurite outgrowth inhibitory factors expressed in and around the lesion. These include several myelin associated inhibitors, proteoglycans, and tenascin-R. Recent studies have documented the presence of class 3 semaphorins in fibroblast-like meningeal cells present in the core of the neural scar formed following CNS injury. Class 3 semaphorins display neurite growth-inhibitory effects on growing axons during embryonic development. The induction of the expression of class 3 semaphorins in the neural scar and the persistent expression of their receptors, the neuropilins and plexins, by injured CNS neurons suggest that they contribute to the regenerative failure of CNS neurons. Neuropilins are also expressed in the neural scar in a subpopulation of meningeal fibroblast and in neurons in the vicinity of the scar. Semaphorin/neuropilin signaling might therefore also be important for cell migration, angiogenis and neuronal cell death in or around neural scars.

In contrast to neurons in the CNS, neuropilin/plexin positive neurons in the PNS do display long distance regeneration following injury. Injured PNS neurons do not encounter a semaphorin positive neural scar. Furthermore, Semaphorin 3A is downregulated in the regenerating spinal motor neurons themselves. This was accompanied by a transient upregulation of Semaphorin 3A in the target muscle. These observations suggest that the injury induced regulation of Semaphorin 3A in

*Netherlands Institute for Brain Research, Meibergdreef 33, 1105 AZ Amsterdam, The Netherlands.

the PNS contributes to successful regeneration and target reinnervation. Future studies in genetically modified mice should provide more insight into the mechanisms by which neuropilins and semaphorins influence nervous system regeneration and degeneration.

INTRODUCTION

Maturation of the mammalian central nervous system is accompanied by a significant decline of its spontaneous capacity to regenerate following injury. In contrast, neurons of the adult mammalian peripheral nervous system (PNS) do retain their regenerative capacity throughout life. The balance between growth-supporting and growth-inhibiting factors, expressed by neurons and non-neuronal cells, is thought to determine whether regeneration occurs successfully or fails.[1-3]

Molecules that inhibit regenerative axonal outgrowth are present in CNS myelin and in the neural scar that is formed at the site of the lesion.[4,5] Following a CNS lesion induction of the expression of growth-inhibitory proteins in the scar is regarded as an important cause underlying regenerative failure in the adult mammalian CNS. The discovery that repulsive axon guidance cues, including members of the ephrin, netrin, Slit and semaphorin family, are also expressed in the mature nervous system[1-3] has led to speculation on possible roles of these proteins in plasticity and regeneration during adulthood.[6-9] Although the levels of these molecules often decline during maturation, many of these proteins and their receptors, are persistently expressed during adulthood.[9-14] The recent reports of injury-induced expression of Semaphorin3A (Sema3A) and Ephrin B3 (EphB3) at the spinal cord lesion site,[15-17] together with the downregulation of Sema3A in motor neurons and the upregulation in terminal Schwann cells after PNS injury,[18-20] strongly suggest that these repulsive axon guidance molecules are involved in neuronal regeneration.

Here we review putative roles of semaphorins and their receptors neuropilins and plexins in the damaged central and peripheral rodent nervous system.

GENERAL FEATURES OF CNS REGENERATION

Following transection of an axon in the CNS, the portion of the axon distal to the lesion starts to degenerate, and will subsequently be removed by macrophages and activated microglia. The injured neuron will either survive and often atrophy when the axon is cut far from the cell body, or die, when the axon is cut near the cell body. At the site of the lesion, a glial scar will be formed. The glial scar consists of two main components. The center of the injury site is invaded by meningeal cells, vascular endothelial cells and macrophages. Around the site of the injury a halo of reactive gliosis, containing astrocytes, oligodendrocyte precursors and microglia cells is formed.[5,21,22] Although most CNS axon populations do form sprouts near the lesion site, these sprouts are not able to grow across the lesion and thus do not re-innervate their distal target cells.[23-27]

Several studies have shown that CNS axons have the capability to regenerate over long distances when provided with a suitable substrate, like a piece of peripheral nerve, Schwann cells grafts, olfactory ensheeting glia cells or fetal nervous tissue.[28-32] This suggests that environmental factors critically contribute to the success of the regeneration process. Successful regeneration of CNS axons into growth-permissive grafts demonstrates their capability to regenerate over relatively long distances. Re-growing CNS axons do stop dead however as soon as they reach the distal end of the graft, and will not reenter CNS tissue. Thus, re-growth of CNS axons through a graft does not lead to the reestablishment of functional neuronal connections.

Further evidence of the inhibitory nature of CNS tissue has been derived from studies on regenerating dorsal root ganglion (DRG) cells. When the central projection of these PNS neurons is crushed between the DRG and the spinal cord, the injured axons will regenerate towards the spinal cord as long as they are in a PNS environment but stop growing as soon as they reach the CNS environment of the spinal cord.[33-35]

To date several inhibitory molecules have been identified in adult CNS myelin, including Nogo[36,37](formerly known as NI-250) and myelin-associated glycoprotein (MAG).[38,39] Antibodies directed to Nogo partially neutralize the myelin-associated inhibition of axonal growth in vitro and in vivo.[37,40-42] Furthermore, CNS injury results in enhanced expression of neurite outgrowth inhibiting extracellular matrix proteins, like chondroitin sulphate proteoglycans (CSPG)[43-46] and tenascin,[47-51] by reactive astrocytes and other scar-associated cells at the injury site.[2,5,22,52] The recent discovery that chemorepulsive proteins, like EphB3[17] and Sema3A,[16,53] known to repel specific developing populations of axons, are expressed at high levels in the injured CNS, provides the first indication that regeneration in the adult CNS might be impaired due to the expression of these repulsive developmental guidance cues.

SEMAPHORIN AND NEUROPILIN IN THE INTACT AND INJURED OLFACTORY SYSTEM

Neuropilin-1 (NRP1) was originally discovered in Xenopus (see chapter 1). Based on its cellular distribution, NRP1 was thought to be an important axon guidance molecule for primary olfactory neurons.[54] Following the discovery that NRP1 is an essential component of the receptor for Sema3A,[6,55] extensive studies have been carried out to relate NRP1 and Sema3A expression to axonal guidance events in the developing, adult and regenerating rodent olfactory system.[9,56-62]

Developing Olfactory System

During early embryonic development olfactory receptor neurons (ORN) extend fibers to the telencephalic vesicle before the formation of their target structure, the olfactory bulb, has started. The arriving ORN axons halt for several days just outside the developing CNS and appear to have a role in inducing the formation of the olfactory bulb.[63] Primary olfactory nerve fibers start to enter the developing olfactory bulb approximately two days after they have arrived at the rim of the telencephalic vesicle.[57,64] Based on the temporal and spatial expression pattern of Sema3A at the periphery of the rat telencephalic vesicle Giger et al[9] suggested the involvement of Sema3A in this pausing behavior of the developing primary olfactory neurites. Renzi et al[58] showed that overexpression of a dominant-negative NRP1 that blocks Sema3A-mediated signaling in primary olfactory axons induces premature ingrowth into the telencephalic vesicle. This study demonstrated that Sema3A indeed governs the pausing of ORN axons at the rim of the telencephalic vesicle.

Analysis of NRP1 and Sema3A expression patterns later on in development of the olfactory system revealed a striking complementary relationship. Growing NRP1 positive sensory fibers avoid Sema3A expressing non-neuronal cells in the nerve layer tracts, resulting in region specific innervation of the olfactory bulb.[61] This is in line with the responsiveness of cultured embryonic ORN to Sema3A.[57,62] In the absence of Sema3A, like in the Sema3A null mutant mice, many NRP1 axon are misrouted, and form atypically located glomeruli.[61]

Sema3A expression in deeper layers of the olfactory bulb, by mitral, periglomerular and tufted cells is thought to prevent overshoot of most primary olfactory axons into extra-glomerular layers of the bulb. [9] However, some primary olfactory axons appear to escape this mechanism and do elaborate transient axons into the external plexiform layer.[65] During embryonic development Sema3A expression is also reported in the olfactory epithelium itself, although its function is not clear.[57,58,62] The relatively low levels of Sema3A found in the embryonic and neonatal olfactory epithelium may push the primary olfactory axons out of the epithelium and towards its target structure, the main olfactory bulb.

Adult Olfactory System

Adult primary olfactory receptor neurons are continuously replaced during adulthood.[66,67] Newly formed olfactory neurons display long distance axon growth towards the olfactory bulb, where they form synapses on their target neurons, the mitral cells and juxtaglomerular cells (Figure 1 and 2). Localization studies showed that NRP1 and collapsin response mediator protein 2 mRNA (CRMP-2; intracellular molecule involved in Sema3A induced growth cone collapse[68]) are predominantly expressed in differentiating and young primary olfactory neurons, suggesting a role for Sema3A in axon pathfinding of these newborn neurons during adulthood[59] (Figure 1A and 1B). In line with this both NRP1 and the cell adhesion molecule L1 (thought to be involved in NRP1 signaling,[69] see chapter 7 of this issue) are present on the primary olfactory axons extending towards the olfactory bulb[64] (Figure 1I). An attractive idea is that Sema3A released by dendrites of mitral and tufted cells may act as a stop signal for growing ORN axons and allows them to establish synaptic contacts within the glomeruli (Figure 1I and Figure 2; box 2). High levels of Sema3A mRNA expression observed in adult mitral and tufted cells and the responsiveness of cultured ORN towards Sema3A during development supports this hypothesis.[56,57,59,60]

Regenerating Olfactory System

During adulthood the primary olfactory system retains a remarkable regenerative response, but ORN axons do not grow through a glial scar and do not reach the forebrain[70] (Figure 1 and Figure 2; box 1). Regenerating ORN axons can, however, reach and enter the forebrain in neonatal mice following bulbectomy. In neonatal mice no inhibiting glial scar is formed.[67] In adult animals, removal of the olfactory bulb induces neurogenesis in the olfactory epithelium[66,71-73] and the formation of a neural scar in the bulbar cavity. The neural scar in the bulbar cavity prevents regeneration of newly formed primary olfactory axons into the undamaged adult CNS.[74,75] Analysis of the CNS scar formed following bulbectomy revealed the presence of high levels of Sema3A transcripts in fibroblast-like cells (Figure 1J and 1K). Encapsulation of the regenerating NRP1 positive axon bundles by these Sema3A positive cells is likely to contribute to the inhibition of their growth thereby preventing them from entering the forebrain.[59]

NEUROPILIN LIGANDS ARE EXPRESSED BY THE FIBROBLAST COMPONENT OF NEURAL CNS SCARS

Like bulbectomy, transection of the lateral olfactory tract (LOT) in the mature brain results in the formation of a neural scar. The LOT contains mainly mitral cell axons which project to the olfactory cortex. Despite the upregulation of growth associated proteins, like B-50/GAP-43,[76] mitral cell axons are not able to grow across the neural scar.[24-27] The cell bodies of the mitral cells, located in the main olfactory bulb, continue to express NRP1 following transection of the LOT, suggesting an ongoing sensitivity of adult injured mitral cell axons for Sema3A. In contrast, NRP1 positive mitral cells do extend axons through the lesion site following LOT lesions in neonatal rats.[16, 24-27] A striking difference between neonatal lesions and lesions during adulthood is the infiltration of numerous Sema3A expressing meningeal fibroblasts that form the core of the adult neural scar. These cells are virtually absent in the neonatal scar.[16,77] Neural scars formed following stab wound lesions into other regions of the adult CNS, such as cerebral cortex, perforant pathway and spinal cord, are all characterized by the infiltration of Sema3A positive meningeal fibroblasts.[16]

Recent examination of the neural scar formed following complete spinal cord transection revealed the expression of all other class 3 semaphorin family members[78] (Figure 3E and 3D). Besides Sema3A, meningeal fibroblasts that penetrate the lesion displayed moderate-to-high expression levels of Sema3B, Sema3C, Sema3E and Sema3F, which are all known to exhibit repulsive properties for subpopulations of axons during neural development.[69,79-81] Continuous expression of the class 3 semaphorin receptor components, NRP1, NRP2 and Plexin-A1 (Plex-A1), by the two major descending motor pathways, the corticospinal tract (Figure 3A-3C) and rubrospinal tract (CST and RST respectively), following spinal

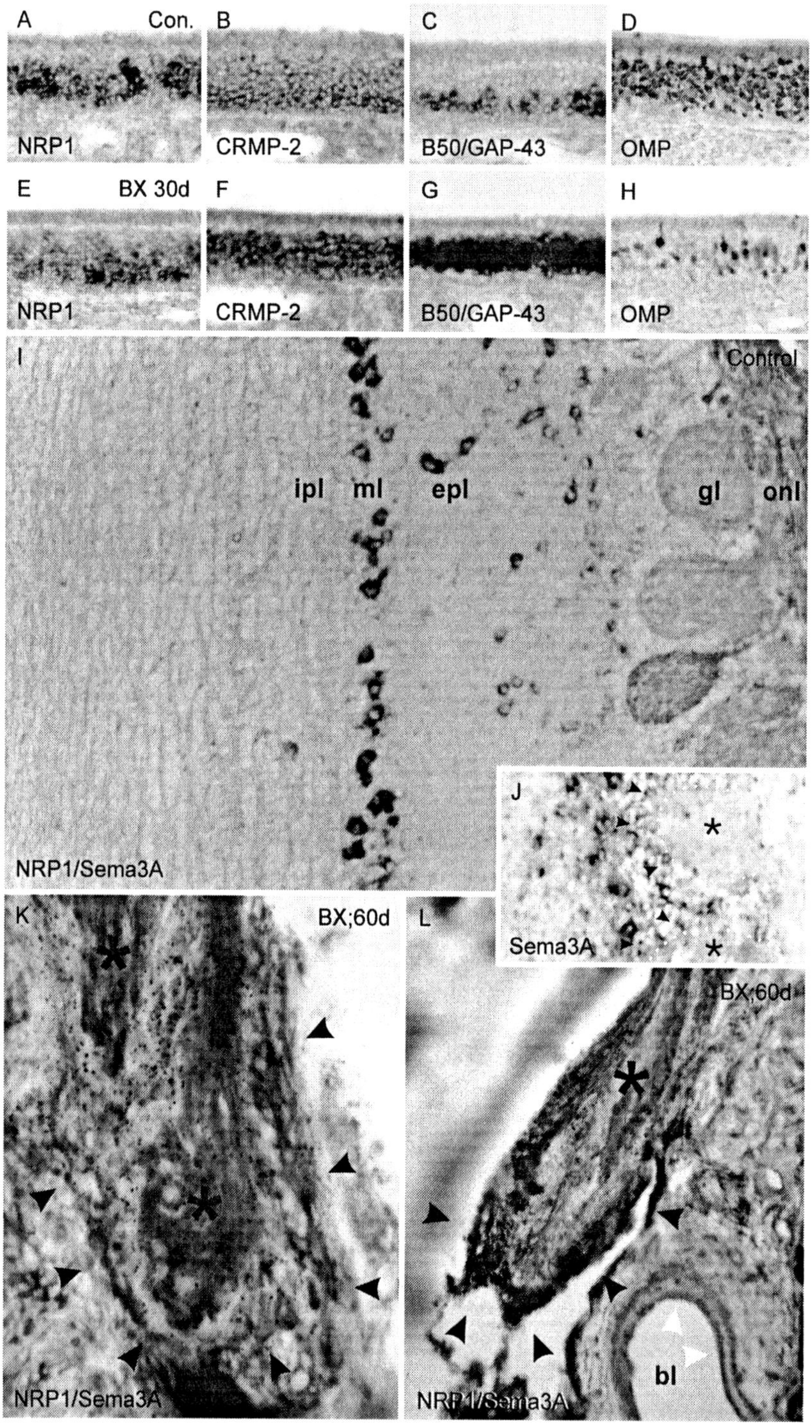

Figure 1. See Legend Next Page.

Figure 1. Neuropilin and semaphorin expression in the intact and injured olfactory system. (A-H) Horizontal sections through the olfactory epithelium of unlesioned adult rats (A-D) and bulbectomized rats (30 days post-lesion, E-F) were stained for NRP1 mRNA (A, E), CRMP-2 mRNA (B, F), B50/GAP-43 mRNA (C, G) and OMP protein (D, H). In control epithelium, NRP1 is expressed by small clusters of ORNs mainly located in the lower and middle parts of the epithelium (A), corresponding to immature B50/GAP43 positive neurons (C) and a subpopulation of mature OMP positive neurons respectively (D). CRMP-2 mRNA is highly expressed in the immature neurons located in the lower regions of the epithelium, but displays a declining gradient toward the apical surface (B). Both sustentacular cells and basal cells, located along the apical and the basal surface respectively, did not display any expression. Following bulbectomy the mature olfactory neurons degenerate and new neurons are formed from cells in the basal cell compartment of the olfactory epithelium. The number of immature B50/GAP-43 positive neurons increases dramatically (G), resulting in a enlarged overlap with the NRP1 (E) and CRMP-2 (F) expressing population. In contrast to the increase in immature neurons is the decrease in the number of OMP positive mature neurons (H), only a few neurons mature without a target. (I-L) Horizontal sections (rostral is to the top) of intact olfactory bulb (I) and 60 days following bulbectomy (K, L) were subjected to in situ hybridization for Sema3A (in purple) and immunocytochemistry for NRP1 (in brown). Mitral cells in the mitral cells layer (ml) and tufted cells located in the external plexiform layer (epl) express high levels of Sema3A mRNA (I). Periglomerular cells, situated on the interface between the external plexiform layer and the glomeruli, express variable levels of Sema3A mRNA (black arrowheads in J). In the internal plexiform layer (ipl) no Sema3A expression was detected. Occasionally small cells in the olfactory nerve layer (onl) showed weak Sema3A expression. NRP1 positive ORN axons grow through the olfactory nerve layer (onl) and terminate in the glomeruli layer (gl) (I). Sixty days following removal of the olfactory bulb, clusters of Sema3A expressing cells (black arrow heads in K and L) encapsulate bundles of NRP1 positive nerve fibers (asterisks) entering the bulbar cavity. A NRP1 positive blood vessel (white arrowheads) in the bulbar cavity is indicated with Bl (L). A color version of this figure is included in the insert that precedes Chapter 1.

cord injury, renders these axon tracts potentially sensitive to scar-derived semaphorins. In line with this, most descending spinal cord fibers fail to enter the semaphorin positive portion of the spinal neural scar. Likewise, Pasterkamp et.al[15] have shown that the ascending central projections of dorsal root ganglion cells also do not penetrate Sema3A positive regions of the scar that is formed after transection of the spinal cord dorsal column.

In summary, NRP1 as well as NRP2 expressing fibers do not penetrate class 3 semaphorin containing regions in the CNS lesion site (Figure 3I). Therefore, injury induced expression of developmentally important chemorepulsive axon guidance molecules, like semaphorins, may contribute significantly to the non-permissive nature of CNS scars.

NEUROPILINS ARE EXPRESSED AT THE CNS LESION SITE

Other functions of class 3 semaphorins in the neural CNS scar should be considered. Recently it has been reported that not only class 3 semaphorins are expressed in the neural scar but also their receptors, the neuropilins[16,53,82] (Figure 3F and 3G). Vascular endothelial growth factor (VEGF), which also binds to neuropilins,[83] is also induced at CNS injury sites.[82,84] The presence of ligands as well as receptor molecules in non-neuronal cells of the CNS scar invites the

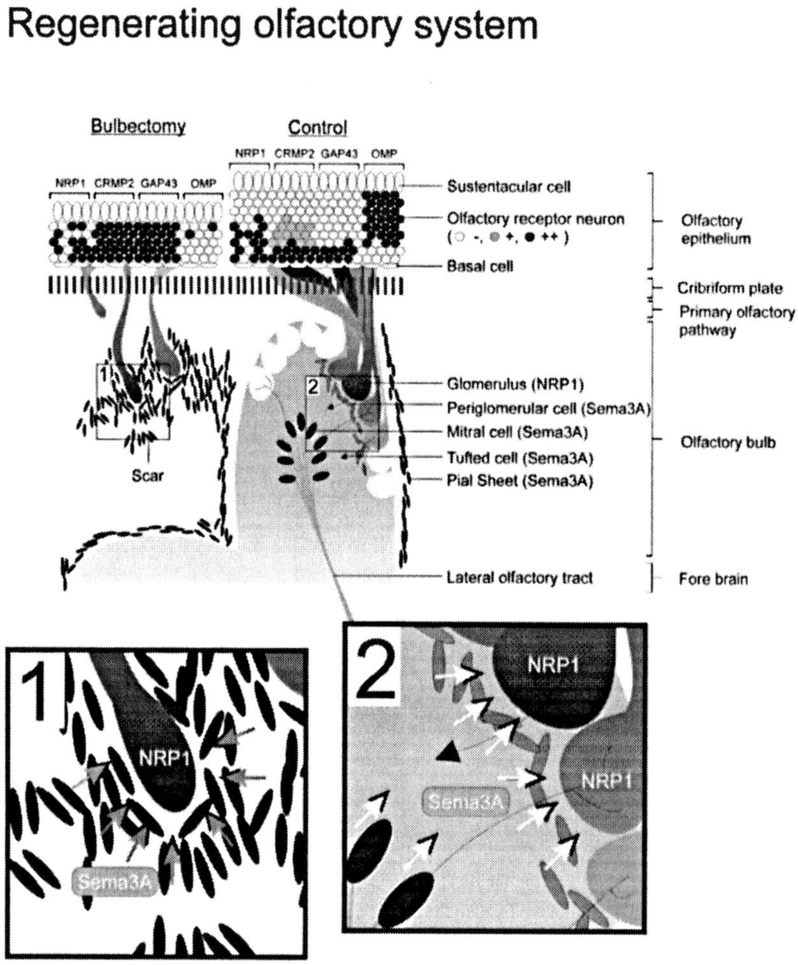

Figure 2. Proposed roles of Neuropilin-Semaphorin 3A signaling in the intact and injured olfactory system. Schematic drawing showing the intact (right side) and the lesioned (left side) olfactory system. Inserts of boxes 1 and 2 are higher magnifications of the corresponding boxes in Figure 2 . ORN are continuously replaced in the intact adult olfactory epithelium. New ORN are derived from the basal cell layer located at the basal side of the epithelium, and migrate in the apical direction during their maturation. Immature ORN express growth-associated proteins, like B50/GAP-43, whereas mature ORNs are characterized by OMP expression. During the time the newly formed ORNs extend their axons through the cribriform plate towards the processes of the secondary olfactory neurons in the main olfactory bulb, they express NRP1 and CRMP-2. Sema3A protein released by dendrites of mitral, tufted and periglomerular cells (white arrows) is thought to serve as a stop signal for the NRP1 positive ingrowing axons (box 2). Forcing them to terminate their extension in the glomeruli layer, and preventing them form growing deeper into the olfactory bulb.Bulbectomy induces an increased turnover of ORNs, visible as an enlarged population of immature, B50/GAP-43 and CRMP-2, expressing cells. Immature ORNs are no longer restricted to the lower regions of the epithelium, but are distributed through the entire epithelium. NRP1 positive axons extended by the immature ORNs grow into the bulbar cavity were they are stopped by Sema3A expressing meningeal fibroblasts (gray arrows) of the neural scar (box 1). Due to the lack of target cells hardly any ORN matures and the majority of the ORNs die prematurely. This results in a decreased epithelial thickness compared to the control side.

speculation that neuropilin-semaphorin/VEGF signaling plays a role in various cellular responses during formation of the neural scar. In vitro studies have revealed that competition between Sema3A and VEGF-165 influences cell survival, migration and proliferation.[85,86]

Migration

In vitro neural crest and endothelial cells are repelled by Sema3A, a process that is mediated via NRP1.[85,87] Sema3C null mutant mice suffer from defects in neural crest cell migration.[88] Furthermore, Sema3C, Sema3E and Sema3F are associated with invasive and metastasizing features of tumor cells.[89-92] It is therefore conceivable that co-expression of neuropilins and semaphorins in the neural scar contributes to cell motility. The temporal expression profile of class 3 semaphorins in meningeal fibroblasts correlates strongly with the massive infiltration of these cells into the scar observed following penetrating injuries of the adult CNS. Following similar injuries in the neonatal brain no semaphorin expression was detected,[16] which is in line with the absence of migrating meningeal fibroblasts in neonatal lesions.[77]

Meningeal fibroblast invasion of the CNS injury site is not only age dependent, but also dependent on lesion type.[5] In non-penetrating injuries, like spinal cord contusion lesions, semaphorin expression is restricted to cells in the swollen meningeal sheet present at the site of the contusion lesion.[78] This can be explained by the limited infiltration of meningeal fibroblasts into a contusion lesion site.[5,93] Although causal evidence remains to be gathered, it is not unlikely that meningeal cell motility during neural scar formation is affected by secreted semaphorins and neuropilins.

Angiogenesis

Studies in genetically manipulated animals demonstrated the importance of neuropilin signaling for blood vessel formation. Both overexpression and absence of NRP1 during development lead to an abnormal cardio-vascular system.[94,95] Additionally, malformations can be observed in the cardio-vascular system of Sema3A knockout mice.[96] Furthermore, Soker et al[83] identified NRP1 as an isoform specific receptor for VEGF-165 (vascular endothelial growth factor-165) which mediates mitogenic activities on endothelial cells. In vitro studies revealed inhibitory effects of Sema3A on endothelial cell motility and microvessel sprouting.[85] It is therefore conceivable that neovascularization observed in and around the CNS lesion area[97,98] may be modulated by injury induced expression of neuropilins, VEGF and class 3 semaphorins.

Incisions in the lateral funiculus of the spinal cord showed that vascular endothelial growth factor and its receptors, VEGFR1 (1fms-like tyrosine kinase 1, Flt-1) and VEGFR2 (fetal liver kinase 1, Flk-1) and co-receptor NRP1 are indeed induced in structures correlated with or near vessels in the lesion area[82] (Figure 3H).

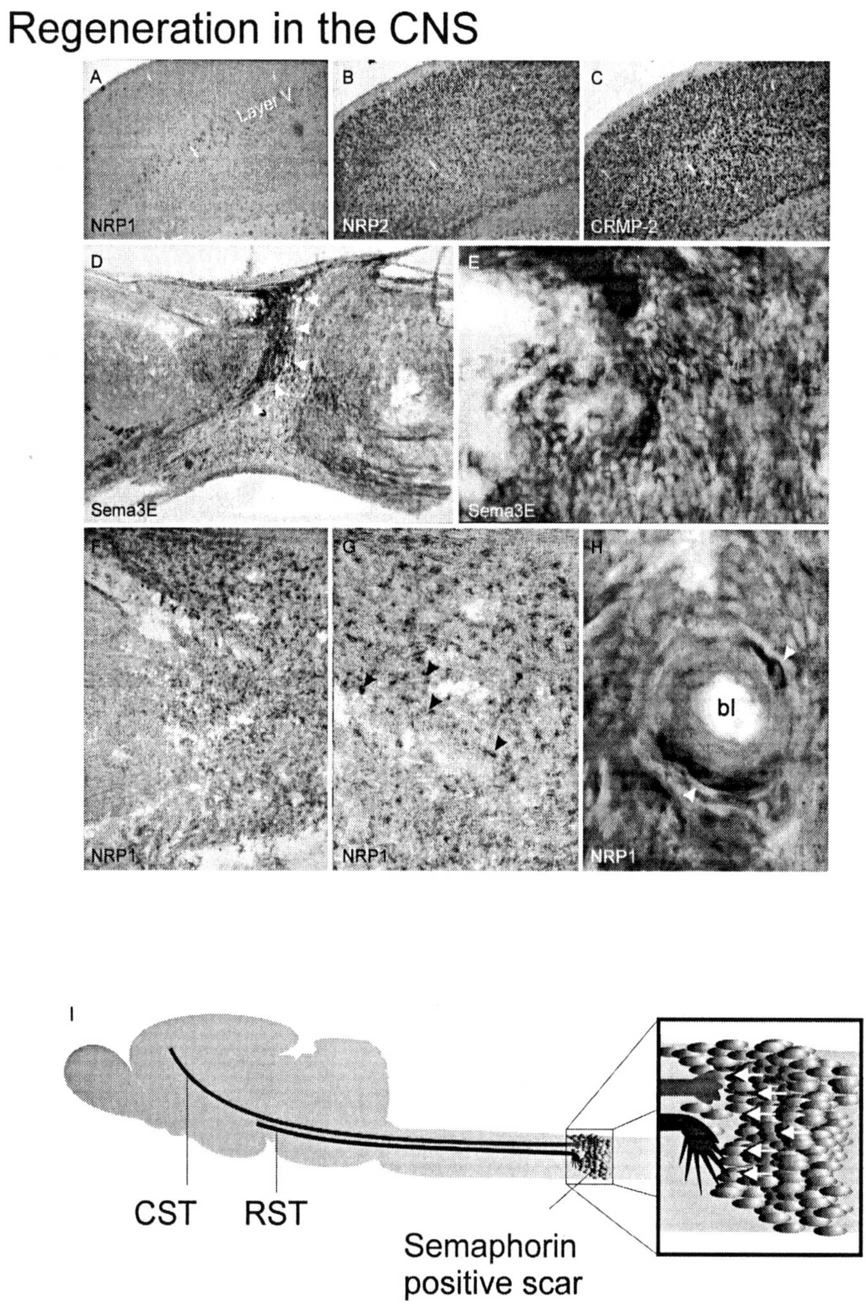

Figure 3. See Legend Next Page.

Figure 3. Neuropilin and Semaphorin expression following complete spinal cord transection. (A-C) Transverse sections through the motor cortex of non-lesioned adult rats were subjected to in situ hybridization for NRP1, NRP2 and CRMP-2. Cell bodies that give rise to the corticospinal tract are located in layer V of the motor cortex (I). NRP1 mRNA is selectively expressed by layer V neurons in the motor cortex (A). Both NRP2 (B) and CRMP-2 (C) are widely distributed through all cortical layers. (D-H) Horizontal sections (rostral is to the left) through the transected rat spinal cord were stained for Sema3E and NRP1 mRNA. At 14 days following transection, clusters of Sema3E positive cells are found deep in the lesion site (D, E is a higher magnification of D). These clusters are organized in strands and are often connected to the meningeal sheet. NRP1 expression is found scattered through the lesion site. (F, G). NRP1 positive cells are often found in close vicinity with blood vessels in the lesion (H). (I) Schematic sagittal overview of the spinal cord transection model. Cortico- and rubrospinal neurons are located in layer V of the motor cortex and the red nucleus respectively, and form the two major descending spinal motor pathways (CST and RST). Following transection of the cortico- and rubrospinal axons, the cell bodies continue to express the messengers for class 3 semaphorin receptor components and the intracellular signaling peptide CRMP-2. This suggests an ongoing sensitivity towards repulsive effects of class 3 semaphorins expressed in the neural scar (box), and might contribute to regenerative failure observed in the CNS.

Pasterkamp et al[16] also observed NRP1 expression on the surface of bloodvessels in the neural scar formed following injuries in other CNS brain areas. Furthermore, focal ischemia induces the formation of new blood vessels mainly in the areas where a temporal upregulation of NRP1 is observed in endothelial cells.[99] In vitro studies have shown that Sema3A and VEGF-165 compete for the NRP1 binding site,[85,86] but to date it is not known if this competition favors blood vessel formation at the CNS injury site.

Cell Death

Injury in the adult CNS often results in secondary cell death in the neural tissue surrounding the lesion. Especially following spinal cord injury, progressive secondary cell death extending to proximal and distal directions, has been observed.[100, 101] Several studies have reported on the participation of neuropilin/semaphorin signaling in processes resulting in cell death.[86,102-104]

Dopamine induced apoptosis in neurons is accompanied by a synchronized induction of Sema3A and CRMP. This can be blocked by antibodies against Sema3A or the receptor NRP1, indicating that Sema3A/NRP1 signaling is involved in apoptosis.[102] Sema3A also induces apoptotic cell death of NGF-dependent sensory neurons.[103] Furthermore, Fujita and colleagues showed that middle cerebral artery occlusion in the adult rat brain induces upregulation of NRP1, NRP2 and Sema3A in neurons of the directly affected area within the three days prior to their death.[104] Progressive secondary cell death is a major problem in (especially) spinal cord regeneration and may be facilitated by neuropilins and semaphorins expressed in and around the lesion site.

NEUROPILIN/SEMAPHORIN REGULATION IN RAT MODELS FOR STATUS EPILEPTICUS

Organotypic cocultures have revealed that embryonic pieces of entorhinal cortex (EC) repel hippocampal axons. This effect can be blocked by NRP1 antibodies or mimicked by Sema3A secreting COS cell aggregates.[105-107] Since neuropilin-semaphorin interactions in the developing nervous system are essential for the formation of a correct neural network,[96,108-110] disturbances here in might be involved in the formation of aberrant neural circuits in the diseased brain. Studies in different rat models for status epilepticus (SE) have revealed changes in neuropilin and/or semaphorin expression prior to axon sprouting and synaptic reorganizations observed in these models.[111-113]

In the temporal lobe epilepsy (TLE) model status epilepticus is induced by electrical stimulation of the axons projecting from EC to the molecular layer (ML) of the dentate gyrus, the so called angular bundle (Figure 4). This results in spontaneous seizures after a latent period of 1-2 weeks which is preceded by sprouting of the granule cell axons (mossy fibers) into the ML. Transient downregulation of Sema3A expression was observed in stellate cells of the EC following induction of SE.[111] Furthermore, GAP-43 was upregulated in the granule cells themselves.[114-117] The loss of a repulsive molecule for mossy fibers and the upregulation of intrinsic growth molecules could allow mossy fibers to penetrate into the ML. Although the lack of Sema3A protein secretion in the OML has as yet not been shown, it is likely to occur since up regulation of GAP-43 by itself is not sufficient to induce mossy fiber sprouting.[111] Axonal sprouting of CA1 pyramidal cells in kainic acid induced SE is not only accompanied by reduction in expression of the ligands Sema3A and Sema3C in these cells, but also by a decline in their NRP1 and NRP2 expression.[112,113]

Induced temporary changes in the expression levels of outgrowth-promoting and outgrowth-restricting molecules may contribute to processes like reactive sprouting in the epileptic hippocampus. Moreover, it suggests the involvement of these molecules in structural plasticity in the intact adult brain. This notion is further supported by the persistent expression of axon growth regulating signaling proteins, like neuropilins, plexins and semaphorins, in adult brain structures, typically associated with plasticity, such as the olfactory-hippocampal system.[56,118]

GENERAL ASPECTS OF PNS REGENERATION

Damage to the adult mammalian PNS is marked by relatively successful regeneration, including functional recovery of motor and sensory functions.[119,120] Transection or crush of peripheral axons results in degeneration of the axon stump and the myelin sheath distal to the lesion. Removal of axonal and myelin debris by macrophages and Schwann cells is essential for successful regeneration.[1,3,121,122] Schwann cells start to divide and initiate the expression of several neurotrophic factors, including NGF and BDNF.[123,124] Furthermore, they change their cell surface by increasing the expression of adhesion molecules, like N-CAM, N-cadherin, and the low affinity receptor (P75) for neurotrophins.[125,126]

The success of regeneration in the peripheral nerve largely depends on the maintenance of basal lamina sheath. Normally these sheaths surround the axon/ Schwann cell units, and are important during regeneration for appropriate guidance of regenerating axons back towards their original targets. Only after severe injury of a peripheral nerve, involving the destruction of the basal lamina and the Schwann cell, a non permissive fibroblastic scar will be formed.

In contrast to axotomized CNS neurons, injured PNS neurons are able to initiate and maintain a gene expression program that promotes axon outgrowth during the time they regenerate. Upregulation of growth-associated proteins, immediate early genes and transcription factors, such as GAP-43/B-50,[120,127] tubulin and actin,[128] c-fos, c-jun and KROX 24,[129-131] are thought to increase the growth potential of lesioned PNS neurons.

NEUROPILIN/SEMAPHORIN REGULATION IN THE INJURED PNS

Facial and spinal motor neurons continuously express NRP1 and Sema3A during adulthood. This indicates that their axons are persistently sensitive to semaphorins, which in turn are continuously present in their vicinity. In contrast to lesions in the CNS, a crush or transection of the peripheral nerve does not induce Sema3A expression at the site of the lesion. In addition, peripheral nerve in (motor) neurons (Figure 5A) while NRP1 and Plex-A1 mRNA levels remain unchanged or are slightly upregulated[18,132] (Figure 5C and5D). The Sema3A messenger levels stay low during the time injured neurons extend regenerating axons towards their target. The period of down regulation is closely related to the temporal upregulation of the growth associated protein B-50/GAP-43 (Figure 5B). Target re-innervation appears to be essential to restore Sema3A expression, since prevention of regeneration by nerve transection and back-ligation of the proximal nerve stump results in persistent downregulation of Sema3A expression.[18]

The biological significance of co-expression of NRP1/PlexA1 complex and Sema3A in the same neuron is currently not understood. One hypothesis states that the regeneration related down regulation of Sema3A is necessary to prevent an inhibitory effect of secreted Sema3A on its own and/or neighboring axon tips. A similar situation is observed in the developing chick embryo, where growing spinal motor neurons express Sema3A, and at the same time are repelled by this molecule in an in vitro assay.[133] Co-culture studies have demonstrated that rat embryonic motor neurons are responsive to Sema3A.[134]

An alternative hypothesis, concerning co-expression of the NRP1/PlexA1 complex and Sema3A in the same neuron, has been formulated based on studies of ephrin. It has been shown that this chemorepulsive guidance molecule can functionally modulate its receptor when co-expressed in the same neuron.[135] If the same is true for Sema3A signaling, regenerating peripheral axons, that have

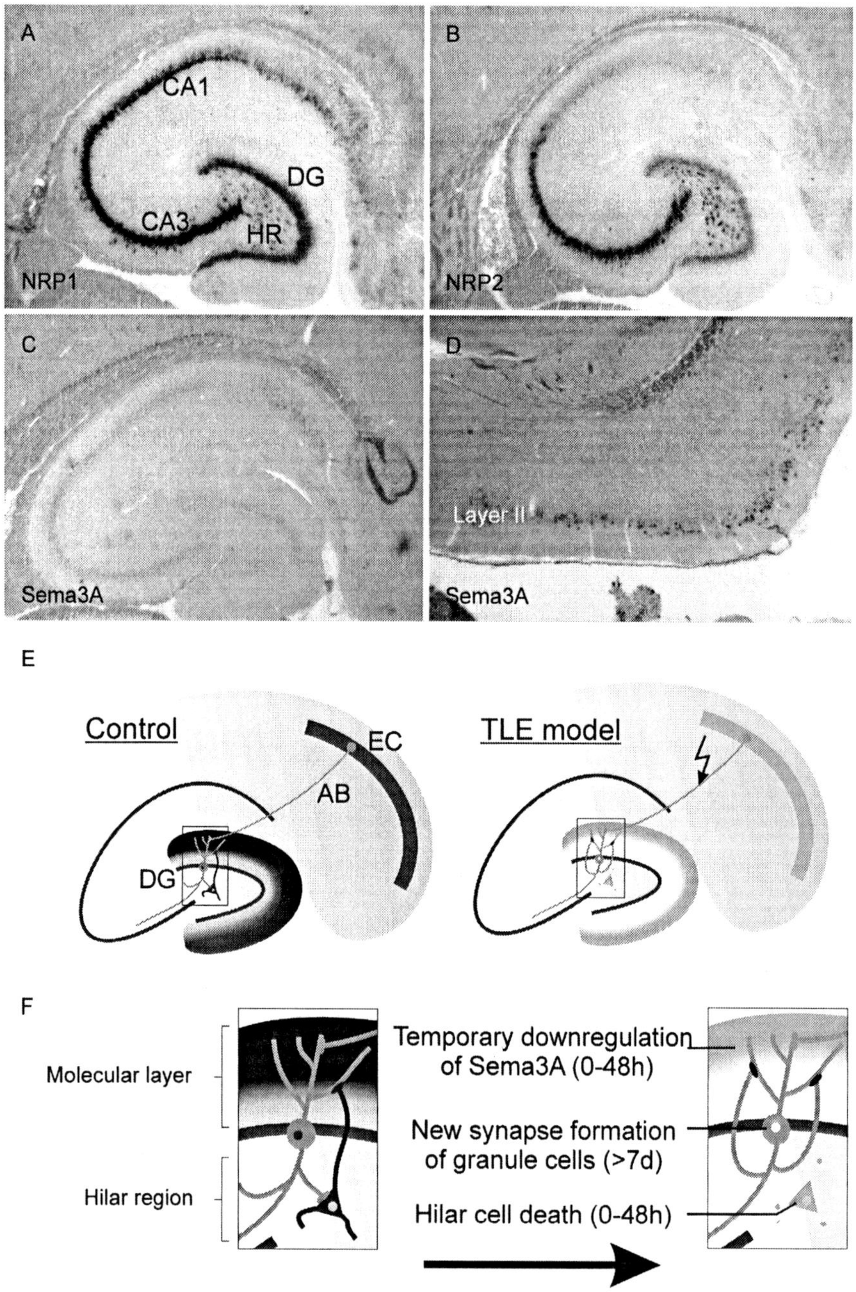

Figure 4. See Legend Next Page

Figure 4. Neuropilin and Semaphorin 3A expression in the entorhinal-hippocampal system. (A-D) Horizontal sections through the adult rat brain. In situ hybridization for NRP1, NRP2 and Sema3A mRNA in the entorhinal-hippocampal system. NRP1 expression is strong in the main structures of the hippocampus, including CA1, CA3 and dentate gyrus, only cells in the hilus have expression. Strong expression for NRP2 was detected in the hilar cell region (HR) and pyramidal cells of the CA3 region of the hippocampus (A). Granule cells in the dentate gyrus (DG) showed moderate levels of expression (A). Stellate cells in layer II of the entorhinal cortex show moderate-to-high Sema3A expression (D). (E) Schematic overview of the proposed role for Sema3A in the rat entorhinal hippocampal system and in the rat TLE model (F shows higher magnifications of the boxed areas in E). Sema3A expressing stellate cells in entorhinal cortex (EC) layer II project their axon via the angular bundle (AB) to the outer molecular layer of the hippocampus. In this area they synapse on the dendrites of the granule cells in the dentate gyrus (DG). Release of Sema3A into the outer molecular layer may contribute to the formation of a repulsive gradient in the molecular layer which normally restricts synaptic reorganization of the NRP1- positive granule cells and hilar cells. Electrical stimulation of the angular bundle results in a temporary downregulation of Sema3A expression by EC neurons, which may result in a transient loss of the chemorepulsive gradient. It further induces death of the hilar cells and induction of GAP43 expression in the granule cells. The temporary loss of a repulsive protein in the molecular layer, together with the increased growth potential of the granule cells, may allow sprouting of the granule cell axons (mossy fibers) into the molecular layer. Where they replace the lost synapses of the dying hilar cells.

downregulated their Sema3A expression, would be more sensitive for Sema3A released from other sources. In this context it is interesting that upon injury, Sema3A expression is induced in terminal Schwann cells at endplates in the target muscle, suggesting a role for Sema3A in post-lesion stabilization of the newly formed neuromuscular junction[19,20] (Figure 5E).

Peripheral nerve injury is not only followed by axon regeneration of the peripheral stump, but also by reorganization of sensory terminal arbors in the dorsal and ventral spinal cord. Among the connections that undergo reorganization are the proprioceptive fibers that synapse on the motor neuron cell bodies and dendrites.[136] Downregulation of Sema3A in motor neurons after nerve injury might contribute to or might even be a prerequisite for altering these and other spinal connections. A subpopulation of sensory neurons in the DRG upregulates or continues to expresses NRP1 following peripheral nerve crush rendering their central projections in the dorsal and ventral spinal cord potentially sensitive for Sema3A.[132] Several studies have shown that both developing and adult sensory fibers are repelled by Sema3A and Sema3E in vitro.[7,79,137-140] Functional evidence that adult neurons in vivo can respond to semaphorins comes from studies in the rabbit cornea. Tanelian et al[8] showed that ectopic expression of Sema3A causes retraction of established, and repulsion of regenerating, Aδ and C sensory fibers in the adult cornea.

CONCLUSIONS

In the injured peripheral nervous system the regulation of semaphorin and neuropilin appears to be consistent with successful regeneration and target re-innervation (Figure 6). Regenerating NRP1/Plex-A1 positive spinal motor neurons do not encounter semaphorins at the lesion site, and even down-regulate their own Sema3A expression. Whether downregulation of Sema3A by the motor neuron itself prevents inhibition of its own axonal growth and/or has a function during the

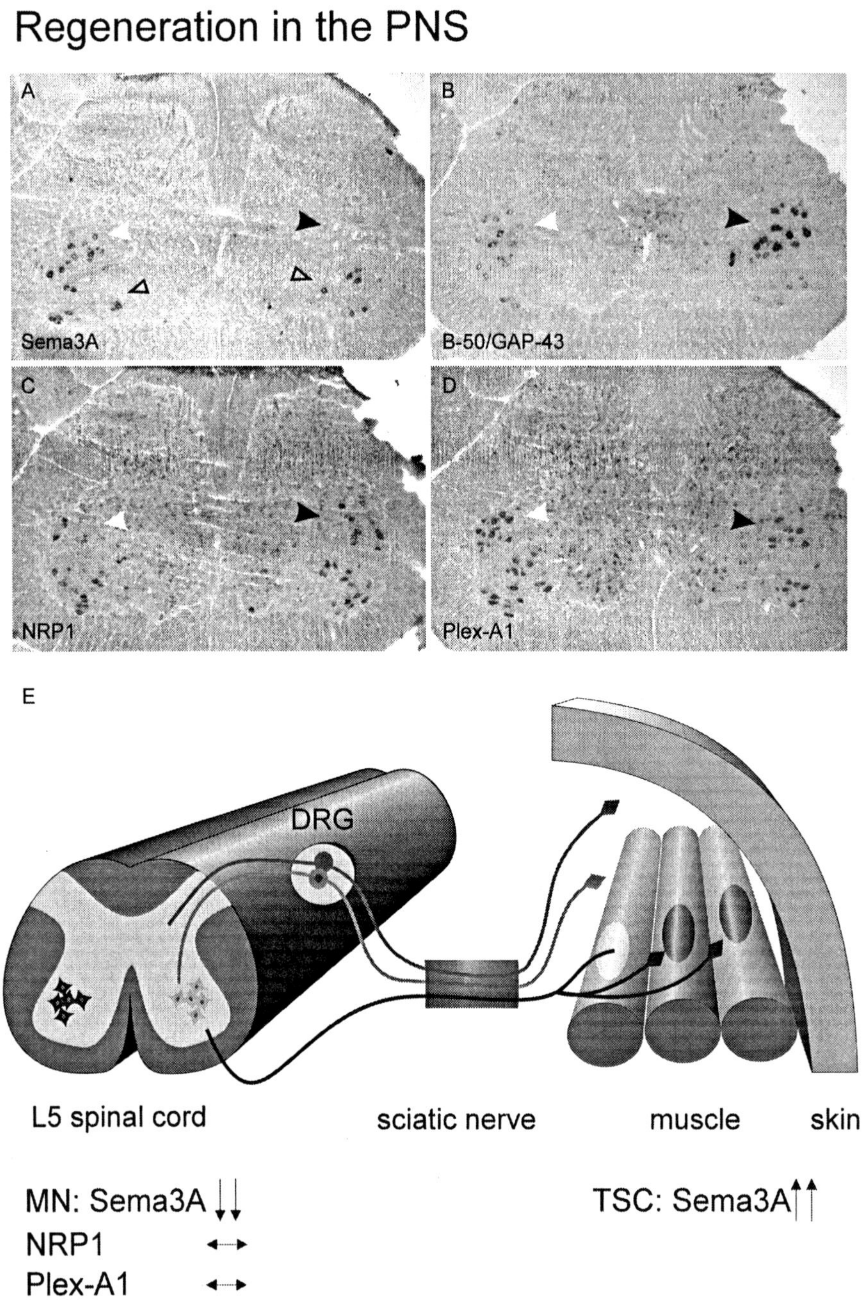

Figure 5. See Legend Next Page

Figure 5. Expression of Sema3A and its receptor components following sciatic nerve crush. (A-D) Serial transverse sections through the rat L5 spinal cord were subjected to in situ hybridization for Sema3A, B50/GAP-43, NRP1 and Plex-A1 mRNA. At 7 days following sciatic nerve crush the lesioned motor neuron pool was identified by high levels of B50/GAP-43 expression (B, arrow head) compared to the low levels controlateral (B, left side). In adjacent sections Sema3A mRNA was not detected in the lesioned dorsolateral pool (A, arrow head), whereas the ventrolateral motor neurons and the motor neurons in the control side continued to express Sema3A mRNA . NRP1 and Plex-A1 were moderately expressed by lesioned and control motor neurons, thus no changes were observed in these receptor genes after PNS injury. (E) Schematic overview of the rat neuromuscular system. Motor neurons located in the lumbar spinal cord project their axons through the sciatic nerve to innervate peripheral targets, like muscle and skin. Following sciatic nerve crush, lesioned motor neurons downregulate the chemorepellent Sema3A, but continue the expression of Sema3A receptor components. This suggests an ongoing sensitivity towards Sema3A derived from other sources then the motor neurons themselves. As a response to denervation, terminal Schwann cells (TSC) in the target muscle start to express Sema3A which might therefore play an important role in the termination of axon growth and target reinnervation in the neuromuscular system.

reorganization of central DRG projections in the ventral and dorsal spinal cord, needs further study. To date, sensory fibers in the rabbit cornea are the only peripheral adult axons proven to be responsive towards ectopically expressed Sema3A.[8]

The appearance of class 3 semaphorins at the adult CNS lesion site correlates with the incapability of adult NRP/Plex positive fibers to penetrate the neural scar. To this date there is no functional evidence elucidating the role of class 3 semaphorins and their receptors in the adult mammalian central nervous system. Future studies should clarify if and how neuropilin/ plexin/semaphorin signaling interferes with CNS regeneration and contributes to various aspects of neural scar formation, including migration and angiogenesis. Inactivation of specific ligands and/or receptors using function blocking antibodies, together with genetic manipulation will provide insights in these distinct roles. Recent studies have shown the possibility to convert the response of growth cones from repulsion to attraction by manipulating the intracellular signaling pathways.[141] This might be a powerful approach to circumvent the inhibitory nature of neural scars and would help to improve the regenerative capacity of the adult mammalian central nervous system.[38]

REFERENCES

1. Fu SY, Gordon T. The cellular and molecular basis of peripheral nerve regeneration. Mol Neurobiol 1997;14(1-2):67-116.
2. Stichel CC, Muller HW. Experimental strategies to promote axonal regeneration after traumatic central nervous system injury. Prog Neurobiol 1998;56(2):119-48.
3. Stoll G, Muller HW. Nerve injury, axonal degeneration and neural regeneration: basic insights. Brain Pathol 1999;9(2):313-25.
4. Horner PJ, Gage FH. Regenerating the damaged central nervous system. Nature 2000;407(6807):963-70.
5. Fawcett JW, Asher RA. The glial scar and central nervous system repair. Brain Res Bull 1999;49(6):377-91.
6. Kolodkin AL, Levengood DV, Rowe EG et al. Neuropilin is a semaphorin III receptor. Cell 1997;90(4):753-62.
7. Luo Y, Raible D, Raper JA. Collapsin: a protein in brain that induces the collapse and paralysis of neuronal growth cones. Cell 1993;75(2):217-27.

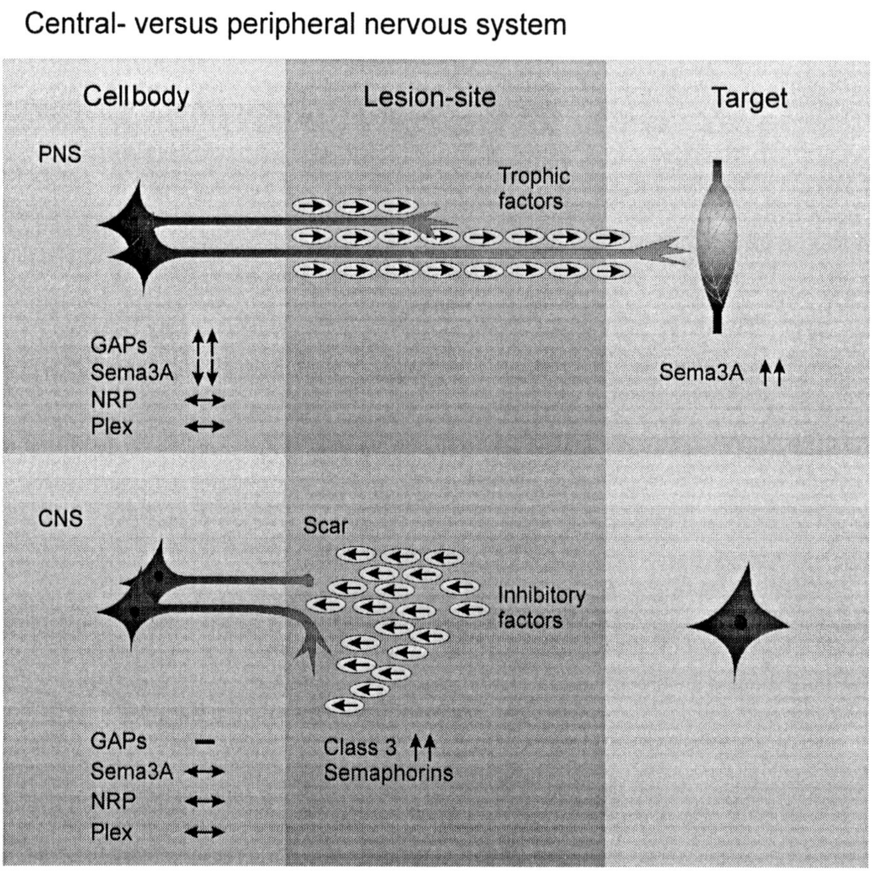

Figure 6. Central- versus peripheral nervous regeneration: proposed role of semaphorins and neuropilins. Peripheral nerve injury induces the expression of a neurite outgrowth supporting gene program (e.g., the upregulation of growth-associated proteins, GAPs) by the injured neurons. Concomitant with upregulation of GAPs the chemorepulsive protein Sema3A is downregulated during the regeneration period while NRP1/Plex-A1 expression levels do not change. Schwann cells actively contribute to a growth-promoting environment, allowing re-growing axons to reach their peripheral targets. Target-derived Sema3A might play an important role in termination of axon growth and target reinnervation. The failure of damaged CNS neurons to initiate a gene program that supports neurite outgrowth, and the formation of a neural scar that contains neurite growth inhibitors (including class 3 semaphorins) results in a persistent denervation of distal targets and permanent loss of function.

8. Tanelian DL, Barry MA, Johnston SA et al. Semaphorin III can repulse and inhibit adult sensory afferents in vivo. Nat Med 1997;3(12):1398-401.
9. Giger RJ, Wolfer DP, De Wit GM et al. Anatomy of rat semaphorin III/collapsin-1 mRNA expression and relationship to developing nerve tracts during neuroembryogenesis. J Comp Neurol 1996;375(3):378-92.

10. Adams RH, Betz H, Puschel AW. A novel class of murine semaphorins with homology to thrombospondin is differentially expressed during early embryogenesis. Mech Dev 1996;57(1):33-45.
11. Livesey FJ, Hunt SP. Netrin and netrin receptor expression in the embryonic mammalian nervous system suggests roles in retinal, striatal, nigral, and cerebellar development. Mol Cell Neurosci 1997;8(6):417-29.
12. Mark MD, Lohrum M, Puschel AW. Patterning neuronal connections by chemorepulsion: the semaphorins. Cell Tissue Res 1997;290(2):299-306.
13. Mori T, Wanaka A, Taguchi A et al. Differential expressions of the eph family of receptor tyrosine kinase genes (sek, elk, eck) in the developing nervous system of the mouse. Brain Res Mol Brain Res 1995;29(2):325-35.
14. Zhang JH, Cerretti DP, Yu T et al. Detection of ligands in regions anatomically connected to neurons expressing the Eph receptor Bsk: potential roles in neuron-target interaction. J Neurosci 1996;16(22):7182-92.
15. Pasterkamp RJ, Anderson PN, Verhaagen J. Peripheral nerve injury fails to induce growth of lesioned ascending dorsal column axons into spinal cord scar tissue expressing the axon repellent Semaphorin3A. Eur J Neurosci 2001;13(3):457-71.
16. Pasterkamp RJ, Giger RJ, Ruitenberg MJ et al. Expression of the gene encoding the chemorepellent semaphorin III is induced in the fibroblast component of neural scar tissue formed following injuries of adult but not neonatal CNS. Mol Cell Neurosci 1999;13(2):143-66.
17. Miranda JD, White LA, Marcillo AE et al. Induction of Eph B3 after spinal cord injury. Exp Neurol 1999;156(1):218-22.
18. Pasterkamp RJ, Giger RJ, Verhaagen J. Regulation of semaphorin III/collapsin-1 gene expression during peripheral nerve regeneration. Exp Neurol 1998;153(2):313-27.
19. De Winter F, Pasterkamp RJ, Verhaagen J. Semaphorin 3A is expressed in terminal Schwann cells during postnatal development and regeneration of adult rat neuromuscular junctions. Soc.Neurosci.abstr. 1999;25:241.
20. De Winter F, Pasterkamp RJ, Stam FJ et al. Injury-induced regulation of semaphorin 3A expression in the neuromuscular system. Soc.Neurosci.abstr. 2000(26):579.
21. Reier PJ. Neural tissue grafts and repair of the injured spinal cord. Neuropathol Appl Neurobiol 1985;11(2):81-104.
22. Reier PJ, Houle JD. The glial scar: its bearing on axonal elongation and transplantation approaches to CNS repair. Adv Neurol 1988;47:87-138.
23. Li Y, Raisman G. Sprouts from cut corticospinal axons persist in the presence of astrocytic scarring in long-term lesions of the adult rat spinal cord. Exp Neurol 1995;134(1):102-11.
24. Devor M. Neuroplasticity in the sparing or deterioration of function after early olfactory tract lesions. Science 1975;190(4218):998-1000.
25. Devor M. Neuroplasticity in the rearrangement of olfactory tract fibers after neonatal transection in hamsters. J Comp Neurol 1976;166(1):49-72.
26. Grafe MR. Developmental factors affecting regeneration in the central nervous system: early but not late formed mitral cells reinnervate olfactory cortex after neonatal tract section. J Neurosci 1983;3(3):617-30.
27. Sijbesma H, Leonard CM. Developmental changes in the astrocytic response to lateral olfactory tract section. Anat Rec 1986;215(4):374-82.
28. David S, Aguayo AJ. Axonal elongation into peripheral nervous system "bridges" after central nervous system injury in adult rats. Science 1981;214(4523):931-3.
29. Aguayo AJ, David S, Bray GM. Influences of the glial environment on the elongation of axons after injury: transplantation studies in adult rodents. J Exp Biol 1981;95:231-40.
30. Richardson PM, Issa VM. Peripheral injury enhances central regeneration of primary sensory neurones. Nature 1984;309(5971):791-3.

31. Vidal-Sanz M, Bray GM, Villegas-Perez MP et al. Axonal regeneration and synapse formation in the superior colliculus by retinal ganglion cells in the adult rat. J Neurosci 1987;7(9):2894-909.
32. Ramon-Cueto A, Cordero MI, Santos-Benito FF et al. Functional recovery of paraplegic rats and motor axon regeneration in their spinal cords by olfactory ensheathing glia. Neuron 2000;25(2):425-35.
33. Carlstedt T. Dorsal root innervation of spinal cord neurons after dorsal root implantation into the spinal cord of adult rats. Neurosci Lett 1985;55(3):343-8.
34. Liuzzi FJ, Lasek RJ. Astrocytes block axonal regeneration in mammals by activating the physiological stop pathway. Science 1987;237(4815):642-5.
35. Bignami A, Chi NH, Dahl D. Regenerating dorsal roots and the nerve entry zone: an immunofluorescence study with neurofilament and laminin antisera. Exp Neurol 1984;85(2):426-36.
36. Caroni P, Schwab ME. Two membrane protein fractions from rat central myelin with inhibitory properties for neurite growth and fibroblast spreading. J Cell Biol 1988;106(4):1281-8.
37. Schnell L, Schwab ME. Axonal regeneration in the rat spinal cord produced by an antibody against myelin-associated neurite growth inhibitors. Nature 1990;343(6255):269-72.
38. McKerracher L, David S, Jackson DL et al. Identification of myelin-associated glycoprotein as a major myelin- derived inhibitor of neurite growth. Neuron 1994;13(4):805-11.
39. Mukhopadhyay G, Doherty P, Walsh FS et al. A novel role for myelin-associated glycoprotein as an inhibitor of axonal regeneration. Neuron 1994;13(3):757-67.
40. Bandtlow CE, Schmidt MF, Hassinger TD et al. Role of intracellular calcium in NI-35-evoked collapse of neuronal growth cones. Science 1993;259(5091):80-3.
41. Caroni P, Schwab ME. Antibody against myelin-associated inhibitor of neurite growth neutralizes nonpermissive substrate properties of CNS white matter. Neuron 1988;1(1):85-96.
42. Tang S, Qiu J, Nikulina E et al. Soluble myelin-associated glycoprotein released from damaged white matter inhibits axonal regeneration. Mol Cell Neurosci 2001;18(3):259-69.
43. Cole GJ, McCabe CF. Identification of a developmentally regulated keratan sulfate proteoglycan that inhibits cell adhesion and neurite outgrowth. Neuron 1991;7(6):1007-18.
44. McKeon RJ, Hoke A, Silver J. Injury-induced proteoglycans inhibit the potential for laminin-mediated axon growth on astrocytic scars. Exp Neurol 1995;136(1):32-43.
45. Snow DM, Lemmon V, Carrino DA et al. Sulfated proteoglycans in astroglial barriers inhibit neurite outgrowth in vitro. Exp Neurol 1990;109(1):111-30.
46. Snow DM, Letourneau PC. Neurite outgrowth on a step gradient of chondroitin sulfate proteoglycan (CS-PG). J Neurobiol 1992;23(3):322-36.
47. Ajemian A, Ness R, David S. Tenascin in the injured rat optic nerve and in non-neuronal cells in vitro: potential role in neural repair. J Comp Neurol 1994;340(2):233-42.
48. Laywell ED, Dorries U, Bartsch U et al. Enhanced expression of the developmentally regulated extracellular matrix molecule tenascin following adult brain injury. Proc Natl Acad Sci U S A 1992;89(7):2634-8.
49. Laywell ED, Steindler DA. Boundaries and wounds, glia and glycoconjugates. Cellular and molecular analyses of developmental partitions and adult brain lesions. Ann N Y Acad Sci 1991;633:122-41.
50. Lips K, Stichel CC, Muller HW. Restricted appearance of tenascin and chondroitin sulphate proteoglycans after transection and sprouting of adult rat postcommissural fornix. J Neurocytol 1995;24(6):449-64.
51. Zhang Y, Anderson PN, Campbell G et al. Tenascin-C expression by neurons and glial cells in the rat spinal cord: changes during postnatal development and after dorsal root or sciatic nerve injury. J Neurocytol 1995;24(8):585-601.
52. Stichel CC, Muller HW. Regenerative failure in the mammalian CNS. Trends Neurosci 1995;18(3):128; discussion 128-9.
53. Pasterkamp RJ, De Winter F, Giger RJ et al. Role for semaphorin III and its receptor neuropilin-1 in neuronal regeneration and scar formation? Prog Brain Res 1998;117:151-70.

54. Satoda M, Takagi S, Ohta K et al. Differential expression of two cell surface proteins, neuropilin and plexin, in Xenopus olfactory axon subclasses. J Neurosci 1995;15(1 Pt 2):942-55.
55. He Z, Tessier-Lavigne M. Neuropilin is a receptor for the axonal chemorepellent Semaphorin III. Cell 1997;90(4):739-51.
56. Giger RJ, Pasterkamp RJ, Heijnen S et al. Anatomical distribution of the chemorepellent semaphorin III/collapsin- 1 in the adult rat and human brain: predominant expression in structures of the olfactory-hippocampal pathway and the motor system. J Neurosci Res 1998;52(1):27-42.
57. Kobayashi H, Koppel AM, Luo Y et al. A role for collapsin-1 in olfactory and cranial sensory axon guidance. J Neurosci 1997;17(21):8339-52.
58. Renzi MJ, Wexler TL, Raper JA. Olfactory sensory axons expressing a dominant-negative semaphorin receptor enter the CNS early and overshoot their target. Neuron 2000;28(2):437-47.
59. Pasterkamp RJ, De Winter F, Holtmaat AJ et al. Evidence for a role of the chemorepellent semaphorin III and its receptor neuropilin-1 in the regeneration of primary olfactory axons. J Neurosci 1998;18(23):9962-76.
60. Pasterkamp RJ, Ruitenberg MJ, Verhaagen J. Semaphorins and their receptors in olfactory axon guidance. Cell Mol Biol (Noisy-le-grand) 1999;45(6):763-79.
61. Schwarting GA, Kostek C, Ahmad N et al. Semaphorin 3A is required for guidance of olfactory axons in mice. J Neurosci 2000;20(20):7691-7.
62. Williams-Hogarth LC, Puche AC, Torrey C et al. Expression of semaphorins in developing and regenerating olfactory epithelium. J Comp Neurol 2000;423(4):565-78.
63. Gong Q, Shipley MT. Evidence that pioneer olfactory axons regulate telencephalon cell cycle kinetics to induce the formation of the olfactory bulb. Neuron 1995;14(1):91-101.
64. Gong Q, Shipley MT. Expression of extracellular matrix molecules and cell surface molecules in the olfactory nerve pathway during early development. J Comp Neurol 1996;366(1):1-14.
65. Santacana M, Heredia M, Valverde F. Transient pattern of exuberant projections of olfactory axons during development in the rat. Brain Res Dev Brain Res 1992;70(2):213-22.
66. Carr VM, Farbman AI. Ablation of the olfactory bulb up-regulates the rate of neurogenesis and induces precocious cell death in olfactory epithelium. Exp Neurol 1992;115(1):55-9.
67. Graziadei PP, Levine RR, Graziadei GA. Regeneration of olfactory axons and synapse formation in the forebrain after bulbectomy in neonatal mice. Proc Natl Acad Sci U S A 1978;75(10):5230-4.
68. Goshima Y, Nakamura F, Strittmatter P et al. Collapsin-induced growth cone collapse mediated by an intracellular protein related to UNC-33. Nature 1995;376(6540):509-14.
69. Castellani V, Chedotal A, Schachner M et al. Analysis of the L1-deficient mouse phenotype reveals cross-talk between Sema3A and L1 signaling pathways in axonal guidance. Neuron 2000;27(2):237-49.
70. Doucette JR, Kiernan JA, Flumerfelt BA. The re-innervation of olfactory glomeruli following transection of primary olfactory axons in the central or peripheral nervous system. J Anat 1983;137(Pt 1):1-19.
71. Graziadei GA, Graziadei PP. Neurogenesis and neuron regeneration in the olfactory system of mammals. II. Degeneration and reconstitution of the olfactory sensory neurons after axotomy. J Neurocytol 1979;8(2):197-213.
72. Graziadei PP, Graziadei GA. Neurogenesis and neuron regeneration in the olfactory system of mammals. I. Morphological aspects of differentiation and structural organization of the olfactory sensory neurons. J Neurocytol 1979;8(1):1-18.
73. Schwartz Levey M, Chikaraishi DM, Kauer JS. Characterization of potential precursor populations in the mouse olfactory epithelium using immunocytochemistry and autoradiography. J Neurosci 1991;11(11):3556-64.

74. Monti Graziadei GA. Experimental studies on the olfactory marker protein. III. The olfactory marker protein in the olfactory neuroepithelium lacking connections with the forebrain. Brain Res 1983;262(2):303-8.
75. Hendricks KR, Kott JN, Lee ME et al. Recovery of olfactory behavior. I. Recovery after a complete olfactory bulb lesion correlates with patterns of olfactory nerve penetration. Brain Res 1994;648(1):121-33.
76. Verhaagen J, Zhang Y, Hamers FP et al. Elevated expression of B-50 (GAP-43)-mRNA in a subpopulation of olfactory bulb mitral cells following axotomy. J Neurosci Res 1993;35(2):162-9.
77. Berry M, Maxwell WL, Logan A et al. Deposition of scar tissue in the central nervous system. Acta Neurochir Suppl 1983;32:31-53.
78. De Winter F, Oudega M, Lankhorst AJ et al. Injury induced regulation of class 3 semaphorin expression in the rat spinal cord. Exp Neurol 2002;in press.
79. Miyazaki N, Furuyama T, Amasaki M et al. Mouse semaphorin H inhibits neurite outgrowth from sensory neurons. Neurosci Res 1999;33(4):269-74.
80. Polleux F, Giger RJ, Ginty DD et al. Patterning of cortical efferent projections by semaphorin-neuropilin interactions. Science 1998;282(5395):1904-6.
81. Polleux F, Morrow T, Ghosh A. Semaphorin 3A is a chemoattractant for cortical apical dendrites. Nature 2000;404(6778):567-73.
82. Skold M, Cullheim S, Hammarberg H et al. Induction of VEGF and VEGF receptors in the spinal cord after mechanical spinal injury and prostaglandin administration. Eur J Neurosci 2000;12(10):3675-86.
83. Soker S, Takashima S, Miao HQ et al. Neuropilin-1 is expressed by endothelial and tumor cells as an isoform- specific receptor for vascular endothelial growth factor. Cell 1998;92(6):735-45.
84. Bartholdi D, Rubin BP, Schwab ME. VEGF mRNA induction correlates with changes in the vascular architecture upon spinal cord damage in the rat. Eur J Neurosci 1997;9(12):2549-60.
85. Miao HQ, Soker S, Feiner L et al. Neuropilin-1 mediates collapsin-1/semaphorin III inhibition of endothelial cell motility: functional competition of collapsin-1 and vascular endothelial growth factor-165. J Cell Biol 1999;146(1):233-42.
86. Bagnard D, Vaillant C, Khuth ST et al. Semaphorin 3A-vascular endothelial growth factor-165 balance mediates migration and apoptosis of neural progenitor cells by the recruitment of shared receptor. J Neurosci 2001;21(10):3332-41.
87. Eickholt BJ, Mackenzie SL, Graham A et al. Evidence for collapsin-1 functioning in the control of neural crest migration in both trunk and hindbrain regions. Development 1999;126(10):2181-9.
88. Feiner L, Webber AL, Brown CB et al. Targeted disruption of semaphorin 3C leads to persistent truncus arteriosus and aortic arch interruption. Development 2001;128(16):3061-70.
89. Christensen CR, Klingelhofer J, Tarabykina S et al. Transcription of a novel mouse semaphorin gene, M-semaH, correlates with the metastatic ability of mouse tumor cell lines. Cancer Res 1998;58(6):1238-44.
90. Yamada T, Endo R, Gotoh M et al. Identification of semaphorin E as a non-MDR drug resistance gene of human cancers. Proc Natl Acad Sci U S A 1997;94(26):14713-8.
91. Martin-Satue M, Blanco J. Identification of semaphorin E gene expression in metastatic human lung adenocarcinoma cells by mRNA differential display. J Surg Oncol 1999;72(1):18-23.
92. Brambilla E, Constantin B, Drabkin H et al. Semaphorin SEMA3F localization in malignant human lung and cell lines: A suggested role in cell adhesion and cell migration. Am J Pathol 2000;156(3):939-50.
93. Li M, Shibata A, Li C et al. Myelin-associated glycoprotein inhibits neurite/axon growth and causes growth cone collapse. J Neurosci Res 1996;46(4):404-14.
94. Kitsukawa T, Shimono A, Kawakami A et al. Overexpression of a membrane protein, neuropilin, in chimeric mice causes anomalies in the cardiovascular system, nervous system and limbs. Development 1995;121(12):4309-18.

95. Kawasaki T, Kitsukawa T, Bekku Y et al. A requirement for neuropilin-1 in embryonic vessel formation. Development 1999;126(21):4895-902.
96. Behar O, Golden JA, Mashimo H et al. Semaphorin III is needed for normal patterning and growth of nerves, bones and heart. Nature 1996;383(6600):525-8.
97. Beggs JL, Waggener JD. Microvascular regeneration following spinal cord injury: the growth sequence and permeability properties of new vessels. Adv Neurol 1979;22:191-206.
98. Imperato-Kalmar EL, McKinney RA, Schnell L et al. Local changes in vascular architecture following partial spinal cord lesion in the rat. Exp Neurol 1997;145(2 Pt 1):322-8.
99. Zhang ZG, Tsang W, Zhang L et al. Up-regulation of neuropilin-1 in neovasculature after focal cerebral ischemia in the adult rat. J Cereb Blood Flow Metab 2001;21(5):541-9.
100. Amar AP, Levy ML. Pathogenesis and pharmacological strategies for mitigating secondary damage in acute spinal cord injury. Neurosurgery 1999;44(5):1027-39; discussion 1039-40.
101. Crowe MJ, Bresnahan JC, Shuman SL et al. Apoptosis and delayed degeneration after spinal cord injury in rats and monkeys. Nat Med 1997;3(1):73-6.
102. Shirvan A, Ziv I, Fleminger G et al. Semaphorins as mediators of neuronal apoptosis. J Neurochem 1999;73(3):961-71.
103. Gagliardini V, Fankhauser C. Semaphorin III can induce death in sensory neurons. Mol Cell Neurosci 1999;14(4-5):301-16.
104. Fujita H, Zhang B, Sato K et al. Expressions of neuropilin-1, neuropilin-2 and semaphorin 3A mRNA in the rat brain after middle cerebral artery occlusion. Brain Res 2001;914(1-2):1-14.
105. Chedotal A, Del Rio JA, Ruiz M et al. Semaphorins III and IV repel hippocampal axons via two distinct receptors. Development 1998;125(21):4313-23.
106. Pozas E, Pascual M, Nguyen Ba-Charvet KT et al. Age-dependent effects of secreted Semaphorins 3A, 3F, and 3E on developing hippocampal axons: in vitro effects and phenotype of Semaphorin 3A (-/-) mice. Mol Cell Neurosci 2001;18(1):26-43.
107. Steup A, Ninnemann O, Savaskan NE et al. Semaphorin D acts as a repulsive factor for entorhinal and hippocampal neurons. Eur J Neurosci 1999;11(2):729-34.
108. Taniguchi M, Yuasa S, Fujisawa H et al. Disruption of semaphorin III/D gene causes severe abnormality in peripheral nerve projection. Neuron 1997;19(3):519-30.
109. Giger RJ, Cloutier JF, Sahay A et al. Neuropilin-2 is required in vivo for selective axon guidance responses to secreted semaphorins. Neuron 2000;25(1):29-41.
110. Chen H, Bagri A, Zupicich JA et al. Neuropilin-2 regulates the development of selective cranial and sensory nerves and hippocampal mossy fiber projections. Neuron 2000;25(1):43-56.
111. Holtmaat AJ, Gorter JA, De Wit J et al. Transient downregulation of Semaphorin(D)III/Collapsin-1 expression in a rat model for temporal lobe epilepsy. Axon guidance & Neural plasticity, Cold Spring Harbor 1998:84.
112. Barnes GN, Luo Y, Ebens A et al. Anatomical and temporal specific patterns of semaphorin gene expression in rat brain after kainic acid induced status epilepticus. Soc Neurosci Abstr 1999;25:1793.
113. Barnes GN, Luo Y, McNamara JO. Parallel regulation of Sema3C and Sema3C receptor expression in CA1 pyramidal cell layer of rat hippocampus after kainic acid induced status epilepticus. Soc Neurosci Abstr 2000;26:575.
114. Benowitz LI, Rodriguez WR, Neve RL. The pattern of GAP-43 immunostaining changes in the rat hippocampal formation during reactive synaptogenesis. Brain Res Mol Brain Res 1990;8(1):17-23.
115. Cremer H, Chazal G, Goridis C et al. NCAM is essential for axonal growth and fasciculation in the hippocampus. Mol Cell Neurosci 1997;8(5):323-35.
116. Bendotti C, Vezzani A, Tarizzo G et al. Increased expression of GAP-43, somatostatin and neuropeptide Y mRNA in the hippocampus during development of hippocampal kindling in rats. Eur J Neurosci 1993;5(10):1312-20.

117. Meberg PJ, Gall CM, Routtenberg A. Induction of F1/GAP-43 gene expression in hippocampal granule cells after seizures [corrected]. Brain Res Mol Brain Res 1993;17(3-4):295-9.
118. Giger RJ, Pasterkamp RJ, Holtmaat AJ et al. Semaphorin III: role in neuronal development and structural plasticity. Prog Brain Res 1998;117:133-49.
119. Skene JH. Growth-associated proteins and the curious dichotomies of nerve regeneration. Cell 1984;37(3):697-700.
120. Skene JH, Virag I. Posttranslational membrane attachment and dynamic fatty acylation of a neuronal growth cone protein, GAP-43. J Cell Biol 1989;108(2):613-24.
121. Brown MC, Perry VH, Hunt SP et al. Further studies on motor and sensory nerve regeneration in mice with delayed Wallerian degeneration. Eur J Neurosci 1994;6(3):420-8.
122. Tona A, Perides G, Rahemtulla F et al. Extracellular matrix in regenerating rat sciatic nerve: a comparative study on the localization of laminin, hyaluronic acid, and chondroitin sulfate proteoglycans, including versican. J Histochem Cytochem 1993;41(4):593-9.
123. Heumann R, Lindholm D, Bandtlow C et al. Differential regulation of mRNA encoding nerve growth factor and its receptor in rat sciatic nerve during development, degeneration, and regeneration: role of macrophages. Proc Natl Acad Sci U S A 1987;84(23):8735-9.
124. Meyer M, Matsuoka I, Wetmore C et al. Enhanced synthesis of brain-derived neurotrophic factor in the lesioned peripheral nerve: different mechanisms are responsible for the regulation of BDNF and NGF mRNA. J Cell Biol 1992;119(1):45-54.
125. Gai WP, Zhou XF, Rush RA. Analysis of low affinity neurotrophin receptor (p75) expression in glia of the CNS-PNS transition zone following dorsal root transection. Neuropathol Appl Neurobiol 1996;22(5):434-9.
126. Gorio A, Vergani L, Ferro L et al. Glycosaminoglycans in nerve injury: II. Effects on transganglionic degeneration and on the expression of neurotrophic factors. J Neurosci Res 1996;46(5):572-80.
127. Benowitz LI, Routtenberg A. GAP-43: an intrinsic determinant of neuronal development and plasticity. Trends Neurosci 1997;20(2):84-91.
128. Bisby MA, Tetzlaff W. Changes in cytoskeletal protein synthesis following axon injury and during axon regeneration. Mol Neurobiol 1992;6(2-3):107-23.
129. Jenkins R, McMahon SB, Bond AB et al. Expression of c-Jun as a response to dorsal root and peripheral nerve section in damaged and adjacent intact primary sensory neurons in the rat. Eur J Neurosci 1993;5(6):751-9.
130. Robinson GA. Immediate early gene expression in axotomized and regenerating retinal ganglion cells of the adult rat. Brain Res Mol Brain Res 1994;24(1-4):43-54.
131. Herdegen T, Blume A, Buschmann T et al. Expression of activating transcription factor-2, serum response factor and cAMP/Ca response element binding protein in the adult rat brain following generalized seizures, nerve fibre lesion and ultraviolet irradiation. Neuroscience 1997;81(1):199-212.
132. Gavazzi I, Stonehouse J, Sandvig A et al. Peripheral, but not central, axotomy induces neuropilin-1 mRNA expression in adult large diameter primary sensory neurons. J Comp Neurol 2000;423(3):492-9.
133. Shepherd I, Luo Y, Raper JA et al. The distribution of collapsin-1 mRNA in the developing chick nervous system. Dev Biol 1996;173(1):185-99.
134. Varela-Echavarria A, Tucker A, Puschel AW et al. Motor axon subpopulations respond differentially to the chemorepellents netrin-1 and semaphorin D. Neuron 1997;18(2):193-207.
135. Hornberger MR, Dutting D, Ciossek T et al. Modulation of EphA receptor function by coexpressed ephrinA ligands on retinal ganglion cell axons. Neuron 1999;22(4):731-42.
136. Woolf CJ, Shortland P, Coggeshall RE. Peripheral nerve injury triggers central sprouting of myelinated afferents. Nature 1992;355(6355):75-8.
137. Messersmith EK, Leonardo ED, Shatz CJ et al. Semaphorin III can function as a selective chemorepellent to pattern sensory projections in the spinal cord. Neuron 1995;14(5):949-59.

138. Puschel AW, Adams RH, Betz H. The sensory innervation of the mouse spinal cord may be patterned by differential expression of and differential responsiveness to semaphorins. Mol Cell Neurosci 1996;7(5):419-31.
139. Pasterkamp RJ, Giger RJ, Baker RE et al. Ectopic adenoviral vector-directed expression of Sema3A in organotypic spinal cord explants inhibits growth of primary sensory afferents. Dev Biol 2000;220(2):129-41.
140. Reza JN, Gavazzi I, Cohen J. Neuropilin-1 is expressed on adult mammalian dorsal root ganglion neurons and mediates semaphorin3a/collapsin-1-induced growth cone collapse by small diameter sensory afferents. Mol Cell Neurosci 1999;14(4-5):317-26.
141. Song HJ, Ming GL, Poo MM. cAMP-induced switching in turning direction of nerve growth cones. Nature 1997;388(6639):275-9.

INDEX

Actin 28, 50, 76, 77, 92, 94, 95, 127
Affinity 4, 17, 18, 60, 63, 68, 71-73, 82-84, 86, 104, 107, 110, 127
Angiogenesis 49, 52, 82, 104, 107, 109, 131
Attraction 27, 28, 29, 31, 91, 97, 98, 131
Axon guidance 1, 6, 13, 17, 21, 28, 29, 49, 56, 60, 81, 92, 96, 97, 111, 116, 121

Cell adhesion 1, 2, 6-11, 61, 78, 91, 92, 97, 118
Central nervous system (CNS) 38, 39, 115, 116, 117, 119, 121, 123, 125, 127, 131, 132
Cerebral cortex 93, 97, 119
Coagulation factor FV/VIII 57
Collapse 39, 56, 61, 62, 65, 71, 72, 74-77, 82, 95, 96, 104, 111
Corticospinal tract 93, 95, 119, 125
COS cells 16, 17, 53, 84
CRMP-2 118, 121, 122, 125
CSPG 117
CST 93, 97, 119, 125
CUB domain 56, 61, 62
Cytoskeleton 50, 53, 77, 86, 92, 94

Dendritic cell 50, 51, 53
Dendritic cells 49-53
DHAND 33, 37, 43
Discoidin domain receptor (DDR) 4, 57
Dorsal root ganglia (DRGs) 10, 22, 39, 96
Drosophila melanogaster 4, 63

Endothelial cell 55, 56, 61, 65, 104, 106, 123
Endothelial cells 6, 33, 34, 49, 52, 56, 65, 66, 82, 87, 104, 107, 116, 123, 125

Fasciculation 9, 13, 21, 25, 34, 38, 92, 97, 103, 104
FV/FVIII (b1/b2) domain 55

GAP43 121, 129
Growth associated proteins (GAPs) 63, 75, 76, 119, 121, 122, 126, 127, 131, 132
Growth cone 27, 53, 56, 61, 62, 65, 76, 77, 91-97, 99, 100, 111, 118
Growth cones 14, 16, 17, 20, 26, 28, 35, 56, 60, 61, 63, 82, 92, 96, 97, 99, 131, 134
GTPases 53, 71, 75-77

Hibridoma 2
Hippocampus 5, 126, 129

IL-4 50

L1-CAM 35
L1-CAM 52

MAb-A5 2-4, 9
MAbs 1, 2
Major histocompatibility complex (MHC) 49
MAM (c) domain 55, 56
Medial longitudinal fasciculus 24
Molecular interaction 1, 4
Monoclonal antibodies 1, 2

Netrin 29, 94, 100, 116
NRP1 3-11, 33-43, 49, 51, 53, 54, 56-67, 91-100, 117-119, 121-123, 125-127, 129, 131, 132
NRP2 4, 5, 7, 8, 33-36, 38-43, 49, 56-62, 65, 66, 125, 126, 129

Olfactory system 5, 23, 117-119, 121, 122

Peripheral nervous system (PNS) 38, 73, 97, 115-117, 126, 127, 129, 131
Plex 1, 2, 4, 8, 9, 56, 63, 65-67, 92, 119, 127, 129, 131, 132
Plexin 1, 2, 4, 13, 15, 35, 53, 55, 56, 60, 63-65, 71-78, 81, 88, 91, 92, 96, 115, 119, 131
Primary immune response 49, 50, 52, 54

Regeneration 5, 115-117, 119, 125-127, 129, 131, 132
Repulsion 19, 21, 23, 27-29, 72, 73, 82, 87, 97, 129, 131
Retinopathy of prematurity (ROP) 42
RST 119, 125

Scar 115-117, 119, 121-123, 125, 127, 131, 132
Sema3A 2, 7, 8, 10, 13, 14, 16-25, 27, 28, 34, 35, 39, 41, 43, 44, 56, 60-63, 65, 66, 67, 71-78, 91-93, 95-100, 104-106, 112, 116-119, 121-123, 125-127, 129, 131, 132
Semaphorin 1, 2, 7, 8, 10, 13-15, 17, 20, 21, 27, 29, 33, 35, 40, 43, 49, 52, 55, 56, 60, 61, 63, 65, 66, 72, 73, 75, 78, 81, 82, 87, 91, 92, 94, 95, 100, 104, 106, 107, 109-111, 115-117, 119, 121-123, 125, 126, 129, 131
Signal transduction 65, 66, 71-73, 83-88
Slit 94, 100, 116
Spinal cord 5, 16, 22-24, 26, 93-97, 116, 117, 119, 121, 123, 125, 129, 131
Sympathetic ganglion (SG) 10, 60, 62

T lymphocyte 50, 51
T lymphocytes 49, 50
Tumor angiogenesis 33, 36, 40, 43, 104
Tumor necrosis factor 33, 37, 50

Vascular endothelial growth factor (VEGF) 1, 6-8, 10, 33, 35-43, 49, 51, 52, 54-56, 61, 64-67, 78, 81, 82, 84, 86-88, 103, 104, 107-109, 111, 112, 114, 121, 123, 125
Vascular injury 33, 43
VEGF receptor 6, 10, 33, 86, 104

Xenopus 1-7, 9, 27, 41, 55, 117